Mathematical Puzzles and Perplexities

Mathematical Puzzles and Perplexities
How to Make the Most of Them

CLAUDE BIRTWISTLE

Distributed in the United States by
CRANE, RUSSAK & COMPANY, INC.
347 Madison Avenue
New York, New York 10017

First Published in 1971

© George Allen & Unwin Ltd 1971

ISBN 0 04 510036 5
72-193923

Printed in Great Britain
in 11 point Times Roman type
by Alden & Mowbray Ltd
at the Alden Press, Oxford

CONTENTS

PREFACE

This book is intended for the general reader who is interested in testing his ingenuity at solving some puzzles, and perhaps at the same time learning a little mathematics, including the topics of the newer mathematics which are finding their way into our schools. It is also intended for those with a more direct interest in mathematics who may wish to find some rather light-hearted problems to help them along. It is not a textbook and there is no intention that the mathematics explained in it shall be more than is necessary to solve the problems clearly and adequately. Those requiring a fuller treatment of any topic should refer to the books listed in the bibliography.

Vast changes have been taking place both in the content and method of teaching mathematics, and the emphasis has been placed increasingly on encouraging mathematical thought in a variety of situations, rather than practising specialized techniques. It is in this spirit that the present book is written. If it can entertain the reader and at the same time stimulate him to some mathematical thought, then the writing—and the reading—of it will have been worth while.

C.B.

NOTE

Where solutions to problems are given in the text,
a note (*Solution follows*) appears at the end of the
statement of the problem, so that readers who
wish to try their hand at a solution may do so
without having first read the solution in the text.
Hints to the solution of other problems will be
found in Chapter 14.

CHAPTER 1

New Readers Start Here . . .

Would you care to check the following addition sum?

$$
\begin{array}{r}
4\ 7 \\
+2\ 5 \\
\hline
7\ 4
\end{array}
$$

(By the way, the answer is correct!)

Or would you like to substitute numbers for letters (one digit for each letter used) so that the following represents a subtraction problem and its answer?

$$
\begin{array}{c}
\textbf{W O R K} \\
\textbf{T H I S} \\
\hline
\textbf{O U T}
\end{array}
$$

Alternatively, what about putting the numbers 1 to 8 inclusive in the squares in the following figure so that the sum of each circle and the sum of each diameter is the same in each case?

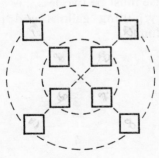

Fig. 1.1

The appeal of a puzzle is universal; most of us respond to the challenge to find the answer to a problem. The three puzzles given above are typical of the large number which appear regularly in books, magazines and newspapers; some are easy, some hard, but all find a ready readership.

However, often there is more to a puzzle than simply finding the answer. After solving the problem, consider first how the answer was found—what deductions are made, what method is used? This enlarges our view from the individual problem to the techniques of solution. Secondly, consider what sort of mathematical knowledge is necessary before one can start the solution, and what mathematics is involved during the solution. This is not merely an understanding of technique, but an appreciation of principle. Finally, is the solution really the end of the problem? Frequently certain aspects arising during the solution are worth further investigation; or the problem or its solution may suggest further investigation of similar situations. (*Solutions follow.*)

The first two points are illustrated in the solutions of the foregoing problems. Normally when faced with an addition as in the first puzzle, we begin by adding seven and five, giving twelve. So we write a 2 in the right-hand (units) column and carry one. Adding this to the second (or tens) column we get a total of seven. Here, however, although our second column total is seven—indicating that one has been carried—the units column has four instead of two.

The secret lies in the different number bases which we use. Normally we count in tens, i.e. our number base is ten. In this case the number in the units column is two more than if we used base ten, so the base must be two less, i.e. eight. If you still do not understand, try putting 'gallons' and 'pints' over the left- and right-hand columns.

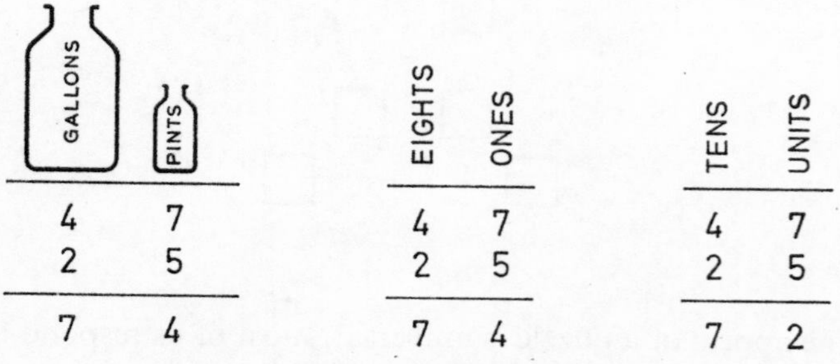

GALLONS	PINTS		EIGHTS	ONES		TENS	UNITS
4	7		4	7		4	7
2	5		2	5		2	5
7	4		7	4		7	2

FIG. 1.2

In solving the second problem we look for clues. T from W is zero, implying that W is one more than T, since they cannot have the same value. This in turn implies that H is bigger than O; yet H from O gives O. Again notice the position of the two Ts. We can now fill in some numbers starting with the hundreds and thousands columns. A solution is

$$2\ 6\ 5\ 4$$
$$1\ 9\ 7\ 3$$

$$6\ 8\ 1$$

The solution to the third problem comes from the realization that the numbers one to eight inclusive add up to thirty-six and that each diameter and each circle contains four of the numbers. Hence the total in each case must be eighteen, and the problem reduces to one of placing numbers appropriately.

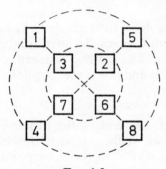

Fig. 1.3

In connection with the second problem, the words '*A* solution . . .' before the answer given are deliberate. Is this a unique solution to the problem or can you find others?

This brings us to the third point made above: when we have obtained one (or more) solutions, need this be the end of the problem? Let us look at a well-known puzzle.

Jean was hard up and decided to go to the pawnbroker's to pawn her gold bracelet which consisted of a chain of seven links. But she didn't want the money all at once; instead she wanted the same amount each day for seven days. For this the pawnbroker said he would take a link of the chain each day as security for the loan. It was necessary, therefore, to cut the chain, but naturally Jean wished to cut as few links as possible. What is the smallest number of links which had to be cut to allow the transaction? (*Solution follows.*)

The answer is fairly easy once one realizes that Jean does not have to leave the same links each day; what she has to do is to leave a given number of links—one the first day, two the second, three the third, and so on. To do this she need only cut one link, the third from one end of the chain. Thus she has one link alone (the cut one), two fastened together, and four fastened together.

FIG. 1.4

The first day she leaves the single link, the second day she collects this link and leaves the two links, the third day she leaves the single link and the two fastened together, the fourth day she collects these and leaves the four links fastened together, and so on.

In a sense this is the end of the problem, certainly of the puzzle as set. But we might extend the problem and go on to consider Kathleen, who is in the same predicament but has a longer chain to exchange for her daily loan. How many links can she have on her bracelet so that two cut links only will give her a maximum number of days' loans? Clearly two cut links can join together three pieces of chain, and what we have to determine is the number of links in each of these pieces in order to allow Kathleen to leave the pawnbroker any whole number of links up to a certain limit.

Consider first the two links which have been cut; these enable Kathleen to leave one and two links but no more. She must, therefore, have a length of chain with three links on it. This, with either or both of the two cut links, enables her to leave up to five links. So her next piece of chain will have to have six links in it, and with these pieces she can leave up to eleven links in all. Hence the final length of chain must have twelve links in it. She can then leave up to twenty-three links in all. Her pieces of chain, therefore, consist of:

> Two cut links
> Three links together
> Six links together
> Twelve links together.

Can you write down, or draw, the different combinations which Kathleen leaves on each of the twenty-three days?

But this is still not the end of the story. So far we have the following maxima:

1 cut will separate a 7-link chain into combinations 1 to 7
2 cuts will separate a 23-link chain into combinations 1 to 23

Suppose we now bring in Lorna and Mary with three and four cuts respectively. Better still, we can deal with both of them (and many more) by taking the general case. Instead of saying three or four, or any other number, let us say n, where n represents any number. The problem then becomes:

We have a chain bracelet in which we are going to cut n links. What is the maximum number of days for which we can make such a chain last? (*Solution follows.*)

If we look at the list of the sections of Kathleen's chain, we see that the first length of chain contains one more than the number of cut links. This is obvious since—using our general case now—if we have n links all separate, we can leave any number up to n. So the first piece of chain we require must have $n+1$ links. This enables us to leave any number of links up to $n+(n+1)$, by using it in conjunction with the n cut links. So the next length of chain must contain one more than $n+(n+1)$, i.e. $n+(n+1)+1$, or $2n+2$ links. (Check this with Kathleen's case by putting $n = 2$).

Now we can cover all possible numbers up to $n+(n+1)+(2n+2)$, i.e. $4n+3$. Hence the next length of chain must have $4n+4$ links. The pattern is now established for the first few terms:

$$n+(n+1)+(2n+2)+(4n+4)$$

and we can write down the succeeding terms either by adding one to the sum of all the terms which have gone before, or by noticing that after the first term each term consists of $(n+1)$ multiplied by a certain constant, thus:

$$n+(n+1)+2(n+1)+4(n+1)$$

The constants are multiplied by two each time, i.e. 1, 2, 4, 8 and so on; this is a geometric series (a later chapter deals with series). It is possible to find the sum of a geometric series by using the easily-proved formula (see Chapter 10)

$$\text{Sum} = \frac{a(r^n - 1)}{r - 1}$$

where a is the first term of the series, r the common ratio (in this case 2) and n the number of terms. But how many terms are there in our series which are multiples of $n+1$? Reference to Kathleen's case will show that there are n such terms. The sum of the numbers 1, 2, 4, 8, ... in our case, therefore, is $2^{n+1} - 1$. The total number of links is then given by the formula

$$n+(2^{n+1} - 1)(n+1)$$

This formula will give us the maximum number of links in the complete chain corresponding to any given number of cuts. Putting $n = 1$, we have $1+(2^2 - 1)(1+1)$, or $1+3.2$, i.e. seven links; this is the solution in Jean's case. The number for Kathleen can be checked by putting $n = 2$. For Lorna, with three cut links, a total of sixty-three links in the original chain enables her to exchange any number of links, one at a time, up to a total of sixty-three days.

More puzzles of this type will be found later in the book.

Next, some problems of a different nature. Of late, the logical type of puzzle has become popular; many of these are particularly wordy and require some careful manipulation of the various statements. Some of the newer mathematical methods can assist in the solution of these and related problems, but such knowledge is not always essential, and a few moments spent in setting out the various statements in tabular form will often produce a solution.

Relatives Mary is the sister of Joan's husband, who is the grandson of Michael's father's brother, and she is engaged to be married to the brother of Joan's sister's husband, whose mother both is Michael's sister, and has an only brother who was Joan's father. What relation was Joan to Mary before her marriage? (*Answer p.* 187.)

We shall give no hint to the solution of this problem, except to say that, in case of difficulty, drawing the family tree might help.

The above problem was one of fixed relationships; some puzzles, however, contain statements which are not fixed, i.e. they may be true or false, and in this case one method of attack is to assess the various possibilities.

Suspects Inspector Brown was trying to find the culprit from his three suspects. He knew that it had been a one-man job, and he knew that each suspect would make one true statement and one false statement.

James said: I didn't do it. Thomson didn't do it.
Thomson said: I didn't do it. Frazer didn't do it.
Frazer said: I didn't do it. I don't know who did.

Which one did Inspector Brown arrest? (*Hint p.* 162. *Answer p.* 187.)

This type of problem sometimes is satisfied by more than one solution. The problem then is not merely to obtain *a* solution, but to obtain all the possible solutions. Examination regulations frequently pose this type of problem, and an easy example is as follows:

Pupil's Choice A candidate wishes to take two science subjects at the GCE O-level examination and may choose from Chemistry, Physics, Biology, Physics-with-Chemistry, General Science and Engineering Science. Due to overlap of syllabuses, however, the following restrictions apply:

Physics-with-Chemistry may not be taken with either Physics or Chemistry.
General Science may not be taken with either Biology or Chemistry.
Engineering Science may not be taken with Physics.
Only one of Physics-with-Chemistry, General Science or Engineering Science may be taken.

B

What possible combinations may the candidate present to make up his two science subjects? (*Hint p*. 162. *Answer p*. 187.)

Harder examples of this type will be found in Chapter 9. The following puzzle is of a type more easily solved by a method of the newer mathematics to be mentioned later in the book, but it is not difficult to find the solution by other methods.

Board Meeting   Four firms had two directors each, and had to send one director to each of three meetings. At the first meeting Archer, Bell, Chivers and Dakin were present; at the second Earnshaw, Bell, Foulds and Dakin were present; and at the third Archer, Earnshaw, Bell and Gerrard were present. Throughout the period Hartley was in bed with mumps. Find the directors of each firm. (*Hint p*. 162. *Answer p*. 187.)

One topic of the newer mathematics is topology. Like so many other branches of mathematics, a thorough treatment of the subject can lead to some very abstruse mathematics, but the elementary part of the subject can be dealt with by experiment and a little basic theory. Essentially this work is concerned with geometry; not the geometry which many of us remember from our school days, but a study of position, relationship, inside and outside, etc. Thus the well-known problem of the Bridges of Königsberg reduces to a problem in topology. Here is an easy start:

Key Situation   A key to a certain door is fastened on my key chain, the other end of which is attached to part of my clothing. I go to the door, unlock it with the key, then—still holding the key in this position—I pass through the door, insert the key and lock the door after me. Is my key chain now twisted or not, and if it is twisted, how many times is it twisted? What happens if I proceed in exactly the same way through a door fitted with a lock on both sides of the door which has a Yale-type key? (*Answer p*. 187.)

The answer may be obtained by experiment, and this is also true of the next problem, but it is a useful exercise to try to work out the answer first.

Carbon Copy I take a sheet of paper, fold it in two so that the bottom half is folded backwards; I then fold it again, but vertically instead of horizontally, and this time the right-hand half is folded backwards. The left-hand and top edges are unfolded and consist of four sheets of paper into which I now interleave three sheets of carbon paper (single-sided carbon and face downwards). I now type or write on the front sheet my name and address. When I have removed the carbon and opened out the sheet again, where and which way up will my name and address appear in its four positions on the paper? (*Answer p.* 187.)

This problem is one of those which may be extended, e.g. suppose I had originally folded the paper, three times instead of twice? Or four times? Five? etc.

The problem illustrates the fact that the idea for a puzzle may arise in many different ways: for example, in the course of one's work or play, which is exactly what happened in the case of the following problem about a typewriter. Those who do not understand the term 'product table', should note that such a table consists of numbers along the top and down the side of the table, the product of any figure along the top and any figure down the side being given where the column and row intersect, thus:

×	2	4	6
3	6	12	18
6	12	24	36

Mixed Product I drew up a product table, then typed it out, but unfortunately all the number keys on my typewriter had been changed. This is what came out:

×	2	3	4	5
2	32	97	12	67
3	97	6	96	98
4	12	96	60	28
5	67	98	28	86

Reconstruct my table. (*Solution follows.*)

It is useful at this point to discuss the method by which this problem is solved, since it has a bearing on other types of puzzle which are to be solved later in the book.

The way into the problem is to study the table generally and notice any particular points which give a lead. Probably the most significant point is the fact that the product of '3' and '3' is '6'. (The inverted commas imply that these numbers are those which appear in the typewritten table.) '6' must represent a single digit square, i.e. 1, 4 or 9. It cannot be 1 since 1^2 is 1, i.e. the product digit would have to be the same as the squared digit; and it cannot be 9 (3×3) because *all* the keys have been changed. Hence '3' stands for 2 and '6' for 4. We may now fill in these figures where they appear in the table, and at the same time notice that the diagonal running from top left to bottom right must consist of square numbers. The rest of the table soon follows:

×	5	2	7	8
5	25	10	35	40
2	10	4	14	16
7	35	14	49	56
8	40	16	56	64

To end this chapter some examples are given of number patterns, which express relationships between numbers. Answers are not given and you are left to discover the patterns yourself. In addition to investigating why these patterns work, you might try your hand at inventing similar relationships. This is not difficult, and a few moments with a pencil and paper should produce quite a few examples.

Complete the following groups of lines, and write down additional lines of the pattern in each case. How far may each pattern be continued?

$$0 \times 9 + 8 = \qquad\qquad 1 \times 9 + 2 =$$
$$9 \times 9 + 7 = \qquad\qquad 12 \times 9 + 3 =$$
$$98 \times 9 + 6 = \qquad\qquad 123 \times 9 + 4 =$$
$$987 \times 9 + 5 =$$

$$3 \times 5 - 4 \times 2 = \qquad\qquad 3 \times 5 - 1 \times 7 =$$
$$4 \times 6 - 5 \times 3 = \qquad\qquad 4 \times 6 - 2 \times 8 =$$
$$5 \times 7 - 6 \times 4 = \qquad\qquad 5 \times 7 - 3 \times 9 =$$

CHAPTER 2

1 and 1 make 10

The first problem of Chapter 1 was an exercise on number bases. Normally the number base to which we work is ten; in other words, we count in tens. The problem set in Chapter 1 was to base eight, so we were counting in eights. There is no particular reason why we should count in tens except that we are more accustomed to this than any other system; most people suggest that our denary system has grown up because we have ten fingers on which to count, and this seems a likely explanation.

In a ten-based system, we have ten digits to form our numbers, i.e. 1 to 9 and 0. This is an important point about all number systems; they contain exactly that number of digits which is the base of the system, and always include 0. Thus a number to base eight would contain any of the digits 1, 2, 3, 4, 5, 6, 7, 0, and a number to base four would contain some or all of the four digits 1, 2, 3, 0.

Secondly, if we write a number, say 235, the right-hand figure (5) denotes 5 units, the next figure to the left (3) denotes 3 multiplied by the base, and the last figure (2) denotes 2 multiplied by the square of the base.

216	36	6	1
	THIRTY-SIXES	SIXES	UNITS
	2	3	5

BASE 6
NUMBER IS 2×36 + 3×6 + 5

	BASE CUBED	BASE SQUARED	BASE	
	n^3	n^2	n	UNITS
		2	3	5

BASE n
NUMBER IS $2n^2 + 3n + 5$

FIG. 2.1

If the base is ten, the number is then, taking it in the order of explanation,

$$5 \text{ units} + (3 \times 10) + (2 \times 10^2)$$

If the number had been in base seven, it would have been

$$5 + (3 \times 7) + (2 \times 7^2)$$

The reason for this is that when we 'carry over' into the next higher column, we are carrying over a ten or a seven or whatever the number is which we are using as base.

The simplest example of all is the binary system, which is to base two; hence there are only two digits, 1 and 0. Below is a simple repeated addition, starting with one and adding one each time. The denary numbers corresponding to the binary totals are shown on the right each time.

```
       1
       1
      ──
      10        2
       1
      ──
      11        3
       1
      ──
     100        4
       1
      ──
     101        5
       1
      ──
     110        6
       1
      ──
     111        7
       1
      ──
    1000        8
       1
      ──
    1001        9        and so on.
```

Taking each successive column below from the right to the left, the digits in those columns represent unity, 2, 4 (2^2), 8 (2^3), 16, 32 and so on.

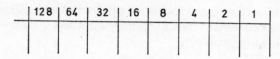

FIG. 2.2

Similarly columns in base 3 would have represented unity, 3, 9, 27, 81, 243, etc.

729	243	81	27	9	3	1

FIG. 2.3

This indicates how we may change numbers in any number system into corresponding numbers in our own denary system. 354 (base 6) will be in the base 10 system

$$4+5\times6+3\times36 = 4+30+108 = 142 \text{ (base 10).}$$

Increased interest has been shown lately in the subject of number bases. In educational circles this has been due partly to the realization that if young children learn to do numerical calculations in various number bases they acquire a better understanding of the principles underlying addition, subtraction, multiplication and division. Another reason for this increased interest—particularly in the binary system—is the increasing use of computers. Computers work by electrical pulses controlled by switches and electronic devices. But a current can only be arranged in two ways—on or off. Hence a corresponding number system has to be used, 1 being equivalent to a switch being on or a current flowing, and 0 being equivalent to the switch being off or no current flowing. Using the binary system we are able, therefore, to feed any number through the computer by a succession of pulses and no-pulses.

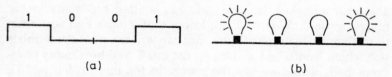

FIG. 2.4 (a) Pulses and (b) lamps represent the number 1001.

Lights of any origin (electric or otherwise) are either 'on' or 'off' and can be used to indicate binary numbers, as, for example, in the following problem.

Advanced Warning To the north of the city state of Nanzipong stands a high mountain, which commands a view of the extensive plain that stretches beyond the northern boundaries of Nanzipong to the regions where barbaric hordes live and plan their attacks on the city state. The elders of the city state make certain that guards are permanently stationed on the mountain, and it is the guards' task when an attack is on the way to light fires to warn the inhabitants of Nanzipong. It is also important to warn the city state of the size of the invading army, which the guards reckon they can do to the nearest thousand. They know that the barbarians number just under fifteen thousand in all. What is the least number of fires which have to be lit in order to convey the required information? (*Hint* p. 163. *Answer p.* 187.)

Earlier we dealt with change of numbers from any base to base ten. Sometimes it may be necessary to change bases when both are not denary. The next problem involves this type of transformation, and is followed by some general problems on number bases.

The Present Xanto lives in the land where they always count to base five and his friend Zimba, lives in the land where they count to base six. The favourite sport of both nations is table-tennis (or ping-pong) and Xanto and Zimba are both keen players. On the occasion of Zimba's birthday, Xanto sent him a present of ping-pong balls, and wrote on the customs declaration form 'This box contains . . . (and here was a three-digit number) ping-pong balls.' When Zimba opened the box and counted the balls he exclaimed 'Xanto has put the wrong number of balls on the box!', and he counted them again. Then he laughed. 'What a foolish mistake!' he said, 'He has written the digits in the reverse order.' Of course, we know that Zimba had forgotten the difference in number bases, but can you say what the number was which Xanto had written on the box? And how many ping-pong balls would *we* say there were in the box? (*Hint* p. 163. *Answer p.* 187.)

Bases and Powers Multiply eleven by eleven in any base greater than two. What do you notice? Remembering this result, can you find a number which is a perfect cube in every base greater than three? Can you continue this process? (*Hint p. 163. Answer p. 187.*)

Elevenses In base ten, write down any number which has an even number of digits. Now reverse the order of the digits, then add the two numbers together. Divide the answer by eleven. Why is this always possible? Will this same property work in other number bases? As an extra exercise, reverting to base ten, write down any number, reverse the digits and then subtract the smaller from the larger number. Find a number which will always divide exactly into the result of this subtraction. (*Hints and Answer p. 163.*)

A Weighty Problem 1 A grocer has a pair of scales of the type with two scale pans where goods to be weighed are put in one pan, and weights to balance the scales are put in the other. What weights are needed in order to be able to weigh any quantity, in pounds weight, up to 40lb, it being assumed that the grocer does not wish to buy more weights than are absolutely essential? (*Solution follows.*)

An inductive approach to this solution seems the best. A 1lb weight is essential. Adding a 2lb weight enables one to weigh up to 3lb. A 4lb weight is the next requirement, and with the other two weights this enables any weight up to 7lb to be obtained. Adding an 8lb weight brings the possibilities to 15lb. By now the pattern should be apparent. Clearly weights of 16lb and 32lb are necessary to round off the grocer's requirement. In fact, this combination of weights will allow the grocer to weigh weights up to 63lb. The relation to the binary system will have been noted.

A Weighty Problem 2 Another solution to the previous problem is possible if the grocer is allowed to place his weights on either side of the scales, i.e. he is allowed to place weights in both scale pans. (*Solution follows.*)

Here the problem involves both addition and subtraction of

weights. Thus if the grocer had a 1lb and a 2lb weight, by placing them on opposite sides of the scales he could obtain a weighing of 1lb. Unfortunately this is no advantage since he can already weigh 1lb with the 1lb weight alone; this is unnecessary duplication. Try 1lb and 3lb weights. Each alone will give 1lb and 3lb; placing them on opposite sides we can get 2lb; and placed on the same side we obtain 4 lb. A much better combination than 1lb and 2lb. Now the pattern can be developed. The first few weighings may be summarized as follows, where + indicates that the weights are added to each other on the same scale pan and − means that they are on opposite sides of the scales:

$$1\text{lb from } 1\text{lb}$$
$$2\text{lb from } (3-1)\text{lb}$$
$$3\text{lb from } 3\text{lb}$$
$$4\text{lb from } (3+1)\text{lb}$$
$$5\text{lb from } (9-3-1)\text{lb}$$
$$6\text{lb from } (9-3)\text{lb}$$
$$7\text{lb from } (9+1-3)\text{lb}$$
and so on.

The solution is that only four weights are needed—1lb, 3lb, 9lb, and 27lb. In the same way that the solution to the first part was related to the binary system (numbers to base two), this result is based on the number system of base three. This problem cannot be generalized in the sense of asking how one uses weights to base 4, 5, . . . , n, but you may care to take the last example (weights allowed on both scale pans) and obtain a formula for the weighings which are obtainable with n weights.

Heads You Win It is required to run a sweepstake by tossing a coin. Thirty-one tickets have been sold. How does one organize this with a minimum number of tosses of the coin, and so that each person has an equal chance of winning? (*Solution follows.*)

The satisfactory solution of this problem depends upon the equal chance requirement. The toss of a coin merely gives a choice between two possibilities and if this involves a choice between an odd number of competititors or alternatives, some competitors are favoured. For example, if we make an original choice between two groups of fifteen and sixteen in the present case, those in the smaller group will have a better chance. In the tournament type of elimination, one gets around odd numbers by arranging 'byes' for certain competitors, but competitors having a 'bye' in any round have a better chance of finally winning than those who have to play a match in that round.

The simplest way to solve the sweepstake problem is to give each competitor a number from one to thirty-one and to toss the coin five times. If it comes down heads, say, we call it 1, if tails we denote it by 0, thereby obtaining five successive figures, 1 or 0, which together give a binary number from 1 to 11111 (or one to thirtyone). This number denotes the winner. Note that there is the possibility of a null result—00000. This means that the five tossings could be used in the same way to find the winner from thirty-two competitors if they were numbered from zero to thirtyone. However, an ordinary tournament-type elimination in reverse (i.e. first toss decides between one group of sixteen and the other group of sixteen; second toss between groups of eight; and so on) also requires just five tosses to find a winner.

Finally, on the subject of number bases, of particular interest are numbers to base twelve. These numbers are called duo-decimals and some people regard twelve as a better base for general everyday use than ten (there is a Duo-decimal Society of Great Britain which holds this view). We shall not discuss here all the arguments for and against, but the easy way in which twelve will split into fractions of a half, a third, a quarter and a sixth is more advantageous than what we find in the case of ten, which can only be divided integrally into halves and fifths (not a very common fraction). It is also claimed that packaging of twelves is easier than tens, which is already evident from the way in which we buy many things in dozens.

One important point about counting to base twelve is that it requires more digits than we have in our number system. You will recall that we said that a number system to base N had $N-1$ digits together with 0; thus the base eight system has the digits 0, 1, 2, 3, 4, 5, 6, 7. So base twelve numbers are going to be made up from zero and eleven more digits; as we only have the nine digits one to nine at present, we shall have to invent two extra digits. There is no standard ruling on this and anyone can use his own notation for these two extra digits. The Duo-decimal

Society use an inverted two, ζ, for the number we normally call ten, and an inverted three, ε, for the number eleven of the denary system. Twelve, of course, is written 10. Hence the digits used in base twelve become 0, 1, 2, 3, 4, 5, 6, 7, 8, 9, ζ, ε. As an example, twenty-three in denary numbers is 1ε in duo-decimal and 4+6 to base twelve is ζ.

Duo-decimals   What denary numbers correspond to the following duo-decimals?

$$792, \quad 3\zeta9, \quad 8\varepsilon\zeta$$

(*Answer p.* 187.)

CHAPTER 3

Patterns in Mathematics

Digital Sum 1 Fit the numbers 1 to 13 in the following spaces in such a way that the sum of the numbers in any straight line is the same. (*Hint p.* 163. *Answer p.* 188.)

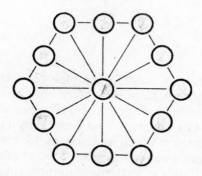

FIG. 3.1

Digital Sum 2 Do the same for the numbers 1 to 12 in the following diagram. (*Hint p.* 163. *Answer p.* 188.)

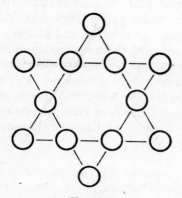

FIG. 3.2

Digital Sum 3 In the same manner, fit the numbers 1 to 9 into the following diagram. (*Hint p.* 163. *Answer p.* 188.)

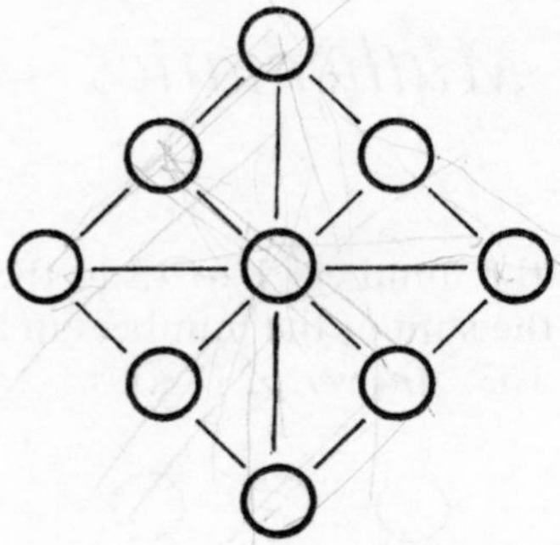

FIG. 3.3

These puzzles are simple examples of patterns made by arrangement of numbers. In some of the examples of the previous two chapters we have discerned other types of pattern. The example in the first chapter about breaking up a bracelet soon established a pattern which enabled us to obtain a formula involving n, where n was any number. The set of examples at the end of Chapter 1 built up patterns of numbers in a different way. In *A Weighty Problem* in Chapter 2 we started by considering the way in which we should use small weights until we saw a pattern in our results, and then we were able to extend it to the solution of the problem.

It is extremely important to look for pattern in mathematics, for pattern signifies order and relationship. Mathematics is the study of relationships in number and in space and by seeking patterns and investigating them we are laying the essential basis of our study. For example, at a very elementary level, the fact that $3+4 = 7$ has no great mathematical significance, but once a child realizes that $3 + 5 = 8$ because 5 is one more than 4 in the first equality and so the answer must be one more than 7, then he is beginning to show mathematical understanding. Take another simple example:

Write down the squares of the numbers 1, 2, 3, 4, . . . and subtract each from the following one. What do you notice about the result? (*Solution follows.*)

$$2^2 - 1^2 = 3$$
$$3^2 - 2^2 = 5$$
$$4^2 - 3^2 = 7$$
$$5^2 - 4^2 = 9$$
$$6^2 - 5^2 = 11$$

and so on.

There are two observations to be made about the pattern that emerges:

(i) we obtain a succession of odd numbers;

(ii) in each case these numbers are the sum of the numbers which we squared.

Why is this?

We can explain it geometrically for any case by drawing a series of squares. Take, for example, $5^2 - 4^2$. Draw a square measuring five units by five units (use squared paper, if possible). Now, within the first square, draw a smaller square of four-unit side.

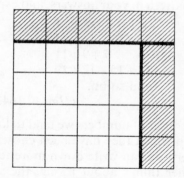

FIG. 3.4

To make the four-unit square into a five-unit square, we have to add two strips, one measuring four squares by one square and the other measuring five squares by one square. The pattern is similar for the other equations, e.g. $4^2 - 3^2$. In each case, to get from the smaller square to the larger square we have to add two strips and these contain the same number of small squares as the sum of the numbers which are being squared. The pattern is the same no matter what squares we use provided that the larger square has a side which is one unit more than the side of the smaller square.

The pattern may also be explained by an algebraic method. If we multiply $(a+b)$ by $(a-b)$ we obtain a^2-b^2. If a is a number which is one more than the number represented by b, $(a-b)$ becomes unity, and we then have simply $a^2-b^2 = a+b$. Giving values to a and b (remember a is always one more than b) the original pattern may be obtained.

The properties used in the above example may be re-arranged to give a different pattern:

$$1 = 1^2$$
$$1+3 = 2^2$$
$$1+3+5 = 3^2$$
$$1+3+5+7 = 4^2$$

and so on.

You should satisfy yourself about the reason for this. (*Hint*: in each case add all the numbers on the left-hand side except the last.)

Odd Sums   As a further exercise, fill in the blanks in the following table, find a pattern in your answers, and explain your result:

$$1 =$$
$$3+5 =$$
$$7+9+11 =$$
$$13+15+17+19 =$$

and so on.

(*Hint p. 164. Answer p. 188.*)

When we see a series of numbers we tend to look for a pattern, i.e. some order in which they have been chosen. Once we have spotted the pattern we can write down more terms of the series and find out other things about it, e.g. the sum of a certain number of terms.

Carry On   Write down the next two terms in each of the following series:

5, 8, 11, 14, . . .
3, 12, 48, . . .
4, 6, 9, $13\frac{1}{2}$, . . .
3, 2, $1\frac{1}{3}$, . . .
1, 1, 2, 3, 5, . . .
a, g, c, i, e, . . .
$\frac{1}{2}$, 1, 2, 5, . . .

(*Solutions follow.*)

The first few examples are simple and the pattern is soon established. In the first, three is added each time, so the next terms are 17 and 20. In the second example, each term is four times the preceeding one, so the series continues with 192 and 768. The third example is similar, but with $1\frac{1}{2}$ instead of 4, and the next terms are $20\frac{1}{4}$ and $30\frac{3}{8}$. The fourth one has each term two-thirds of the one before it and this has the effect of making successive terms diminish; the answer is $8/9$ and $16/27$. The last three are more difficult. In the first of these each term is obtained by adding the two preceeding terms, and the required answer is 8 and 13. The next example is a simple counting forwards and backwards of letters of the alphabet: six forward and four back. In this case the next terms are k and g. The final example shows how difficult it can be to spot a simple pattern when one is not given any indication of its origin. Searching for a relationship between the numbers—as in the previous cases— yields no result; in other words we cannot establish the pattern. In fact, the terms in this case are the number of new pence in successive British coins, i.e. $\frac{1}{2}$p, 1p, 2p, 5p, so the next coins are 10p and 50p and the numbers which follow next in the series are 10 and 50.

In later chapters there will be more problems involving number relationships, although sometimes this may not be stated. The search for pattern should be spontaneous because it brings meaning to what would otherwise appear to be a random display of numbers.

Another type of mathematical pattern concerns geometrical shapes, which is what most people think of when they talk about pattern. But again, the mathematician is not content to look at patterns and to make patterns; he wants to find out how they are formed. As an interesting example of this, certain mathematicians have analysed wallpaper patterns and have noticed the different ways in which the patterns repeat (vertically, horizontally, reflected, and so on). From this work they have shown that there are only seventeen basic ways in which a wallpaper pattern can be repeated.

Similar examples of the way in which we can classify patterns and find order and limitations in them are present in surface covering and space-filling work. Surface-covering patterns are familiar to us in the form of tiled floors, walls and ceilings.

C

Most tiles are square or rectangular in shape and any patterns in the work usually arise from the use of different coloured tiles. In mathematics, on the other hand, we are more concerned with the patterns which may be produced by using different shapes, and we call this work the study of tesselations, which simply means tiling patterns.

In doing the following problems you may find it useful to cut shapes out of cardboard which you can then use as templates to draw around with a pencil. In this way time can be saved in ruling lines and measuring angles, and shapes can be fit together and patterns built up by moving the template around or turning it over, and fitting it into the pattern already drawn.

Rectangular Tiling   As an easy start, take a rectangle whose length is three times its width and see how many different patterns you can form with such tiles. (*Solution follows.*)

The simplest way is to arrange our rectangles in rows and columns.

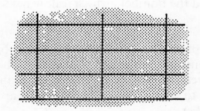

FIG. 3.5

This leads on to the 'staggered' pattern that one sees in brick walls.

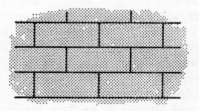

FIG. 3.6

In the example shown above the pattern repeats itself in

position in alternate rows, because the lateral shift from one row to the next is half a rectangle. This could be any fraction, of course; can you formulate a rule about how many rows will be needed before the pattern repeats itself for any given lateral shift?

It is not essential that we should arrange our rectangles in rows, and a zig-zag pattern can be made like this:

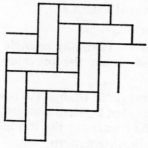

FIG. 3.7

We often see this pattern in wood parquet flooring. The fact that our rectangles have their lengths as a multiple of their widths means that we can build these up into squares by putting three tiles alongside each other.

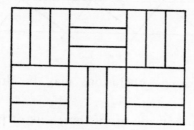

FIG. 3.8

This, too, is an example often seen in parquet flooring. In the pattern above the squares are arranged in horizontal and vertical rows, but we may, if we wish, stagger these in the way that we staggered the individual rectangles in our first pattern above.

We see, then, that although there are certain basic arrangements, there is really no limit to the number of pattern designs which we may make with our rectangular tiles.

Triangular Tesselation 1   Take an equilateral triangle (i.e. all three sides of equal length), and build up a pattern over a fairly large area.

1. How many different patterns can you build?
2. When you have a pattern where corners of six triangles meet at a point, what other regular shapes do you see in your overall pattern? (A 'regular' figure is one which has all its sides of equal length.)

(*Answer p.* 188.)

Triangular Tesselation 2   Take any triangle whose sides are of different lengths and repeat the previous exercise. (*Solution follows.*)

It is possible to arrange the triangles in rows and stagger these rows in different ways, but we shall concern ourselves only with those cases where the corners of the triangles meet in a point. We then have the following pattern

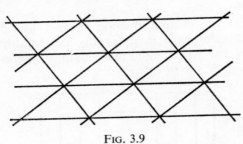

Fig. 3.9

Certain interesting conclusions may be drawn from the resulting figure:

1. The pattern is made up of three sets of parallel straight lines.
2. It becomes clear that we may tesselate a plane surface using a triangle, a parallelogram or a hexagon.
3. These figures—triangle, parallelogram, or hexagon—need not be regular, but in the case of the hexagon and parallelogram the opposite sides are equal and parallel (this is always a property of a parallelogram).

Conclusion 3 above can lead to further investigation as suggested in the following examples:

Quadrilaterals Is it possible to tesselate a plane surface using any quadrilateral (i.e. a four-sided figure with sides of different lengths)? (*Answer p.* 189.)

Tiling Shapes Is it possible to tile a surface with tiles of the shapes set out in each of the examples below? In each case give a mathematical reason for your answer, and try to draw a general conclusion about the necessary condition(s) for a given shape to form a satisfactory tiling unit.

(a) A regular five-sided polygon (i.e. a pentagon)
(b) A regular six-sided polygon (i.e. a hexagon)
(c) A six-sided polygon with sides of different lengths.
(d) A regular eight-sided polygon (i.e. an octagon).

(*Answer p.* 189.)

Multi-shaped Tesselations Is it possible to tesselate a surface in a regular pattern (i.e. a pattern repeating at regular intervals) using the following combinations of figures? (Draw the patterns if possible.)

(a) Regular octagons together with squares whose sides are the same length as the sides of the octagon.
(b) Regular twelve-sided polygons together with equilateral triangles whose sides are the same length as the sides of the polygons.
(c) Regular hexagons together with equilateral triangles whose sides are the same lengths as the sides of the hexagon.
(d) Regular hexagons together with equilateral triangles whose sides are the same lengths as the sides of the hexagons, *and also* squares whose sides are the same lengths as the sides of the hexagons.

(*Hints & solution* p. 164.)

The last example seems to suggest that we can go on making up combinations of regular figures into an infinite number of tesselations, but certain limits do exist. The number of different tesselations which can be used to cover a surface, using regular figures only and such that at each vertex there shall be a meeting of the same number of polygons of the same shape as those meeting at any other vertex, is only eleven. Of these tesselations, only three (square, triangle and hexagon) consist of one figure, i.e. a tile of one regular shape and size. The proof of these statements need not concern us here, but it is interesting to notice the way in which the mathematician again brings order into pattern and draws a conclusion.

x and n—The Unknown and the General

In this chapter we are concerned with problems based on simple algebra and arithmetic. As numerical problems are made up from the ten digits, a problem on these is appropriate.

Century Ahead Arrange mathematical signs in this row of digits

0 1 2 3 4 5 6 7 8 9

so that the answer is exactly 100. (*Answer p.* 189.)

The basic method of solution of an algebraic problem is to put a letter, x say, for the number which we are trying to find, then to obtain an equation which, on being solved, gives the value of x and the solution of the original problem. The main source of difficulty for beginners lies in getting the correct equation involving the unknown quantity, and a common error is to be in too much haste to get this equation. It is advisable to write down a number of statements involving the unknown before finally connecting them in an equation. The working of the following example illustrates this.

A Boy and his Dog Bob's age in years is twice the age of his dog in months, and in two years' time the dog's age in months will be two more than twice Bob's age in years. Find Bob's age now. (*Solution follows.*)

Let x years be Bob's age now. Then his dog's age is $\frac{1}{2}x$ months.

In two year's time Bob's age will be $x+2$ years, and his dog's age will be $\frac{1}{2}x+24$ months.

Hence

$$(\tfrac{1}{2}x+24)-2 = 2(x+2)$$
$$\tfrac{1}{2}x+22 = 2x+4$$
$$1\tfrac{1}{2}x = 18$$
$$x = 12$$

Bob's age now is 12 years.

Most elementary algebra books contain lots of examples of this type, but those given below may need more initial thought. The method of solution, though, is similar to that given in the example above.

John and Mary   John is now twice the age that Mary was when she was five years older than her age when she was a quarter of the age that John is now. How old is John? (*Hint p.* 165. *Answer p.* 189.)

Partners   Ames and Benson were partners in a firm, Ames having one and a half times as much capital invested as Benson. Charles wishes to join the firm and invests a capital of £2500 in it. It is shared between Ames and Benson in such a way that the three partners have equal shares. How is it shared? (*Solution follows.*)

Often a problem which can be solved by using one variable is made somewhat easier by using two variables. In the present example, the method using one variable is given in the 'Hints', (p. 165) but a method with two variables is given below.

Let £x be Benson's original investment.
Then £$1\tfrac{1}{2}x$ is Ames' original investment.
Let £y be the amount Benson receives from Charles.
Then Ames receives £$(2500-y)$ from Charles.

Benson now has invested £$(x-y)$ and Ames has £$1\tfrac{1}{2}x-(2500-y)$. Since they all now have equal amounts invested, Ames and Benson must have £2500 each invested. Hence

$$x-y = 2500 \qquad (1)$$

and

$$1\tfrac{1}{2}x-(2500-y) = 2500$$
$$3x+2y = 10\,000 \qquad (2)$$

Adding twice (1) to (2)

$$5x = 15\,000$$
$$x = 3000$$

and

$$y = 500$$

So the £2500 should be shared so that Benson receives £500 and Ames receives £2000.

Algebra is not always essential for a solution, and we can solve this problem without its use. At the end Ames and Benson must have £2500 each invested. But between them they have received Charles' £2500. Hence their total initial investment must have been £7500. Of this we know that Ames has $1\frac{1}{2}$ times as much invested as Benson, so that Ames must have had £4500 and Benson £3000. It follows that Ames will have to receive £2000 and Benson £500.

Joan and Bill   Bill is twice the age that Joan was when Bill was Joan's age, and when Joan is as old as Bill is now, their total ages will be 63; what are their present ages? (*Hint p.* 165. *Answer* p. 189.)

Trenching   A man is digging a trench of constant width and with a constant downward slope at the bottom (i.e. the trench is getting deeper at a steady rate). When he measures along the level surface the length that he has dug, he finds that he has only extended the trench half the length in the second hour that he had done in the first hour. Assuming that he removes the earth at a constant rate, how much deeper is the end of his trench than the depth at the beginning? (*Hint p.* 165. *Answer* p. 189.)

The method used in solving problems is obviously of great importance, and there is often more than one approach (as has been seen in the case of *Partners* above). Usually the most concise proof is to be preferred, both from the point of view of the time involved and for ease of understanding. A case in point is the following problem.

Knock-out   A knock-out competition has been arranged in some sport. This is the sort of tournament which one sees in the FA cup in soccer, or in a tennis tournament. The players or teams are arranged in pairs, the winner from each pair meeting the winner from another pair, and so on, until an overall victor emerges from the final tie, e.g.

Team A
Team B
Team A
Team C
Team D
Team C
Team C
Team E
Team F
Team F
Team G
Team H
Team G
Team F
Team C

Assuming that there are no byes in the fixture list, that there are no draws, and that n teams initially enter the competition, how many matches must take place before the final winner emerges? (*Solution follows.*)

In this problem the temptation is to start at the beginning by saying that since there are n initial entries, the first round must consist of $\frac{1}{2}n$ matches, and then find the number of matches in successive rounds, the aim being to find the total number of rounds and hence the total number of matches. Careful initial thought, however, shows a more elegant method of solution. Since there is only one ultimate winner, and since each match results in one player (or team) being defeated, and *always* one player for each match, the $n-1$ players who lose must have been eliminated in $n-1$ matches, which, therefore, must have been the number of matches played.

Fly Catcher A train starts from station A to go to Station B 160 miles away at the same time as a train starts from B to go to A. Ignoring the initial acceleration, the first train is travelling at 35 m.p.h., while the train from B to A is travelling at a constant 45 m.p.h. As the trains start to move, a fly leaves the engine of the first train and flies at 60 m.p.h. along the track towards the second train. Upon reaching the engine of the second train, the fly immediately travels back along the track at the same speed until it reaches the first engine, and so on, until the trains meet in a head-on crash which kills the fly. How far does the fly travel in all? (*Hint p.* 166. *Answer p.* 189.)

Some general problems follow.

Birthday Party   The birthdays of a father and his three sons all fall on the same day. On his sixtieth birthday their combined ages will be one year less than his age, and the ages of two of them will be such that one is twice the age of the other. On his eightieth birthday, the combined ages of the same two will be one less than his age. What will be the ages of the three sons when their father is sixty? (*Hint p.* 166. *Answer p.* 189.)

Procession   I was standing at the side of the road watching a procession. It was $1\frac{1}{2}$ miles long and passed me in $\frac{3}{4}$ hour. As the head of the procession passed me, a marshall set off to go to the rear of the procession. Upon reaching the rear, he turned round immediately and returned to the head of the procession, travelling at the same speed throughout. He passed me on his return journey when just half the procession had passed my point. What was his speed? (*Hint p.* 167. *Answer p.* 189.)

Worth a Note   I have a certain number of £1 notes, £5 notes and £10 notes in my wallet. I find that the value of the £5 notes is equal to the value of the £10 notes, and in order to have more £1 notes than £5 and £10 notes combined, I would need to have at least twice the present number of £1 notes. I also notice that if I changed into £1 notes the same number of £5 notes as I have £10 notes, the value of my £5 notes would then be three-quarters of the value of my £1 notes. How many notes have I altogether in my wallet? (*Hint p.* 167. *Answer p.* 189.)

The solution of equations is not always straightforward, as any mathematician will tell you. There are times when an equation may only be solved by using a special technique; sometimes only an approximate solution may be obtained; other equations may give no *definite* solution, but an infinite number of possible solutions; finally, of course, there are those equations which yield no satisfactory solution at all.

We shall consider one or two of these special cases, the first being a fairly well-known problem. There are several versions of this problem using different dimensions, but these variations make little difference to the method of solution.

The Ladder and the Box Problem A ladder is ten feet long and rests with its foot on the ground and its top against a perpendicular wall. In the corner, between the floor and the wall, is a box, measuring three feet high and three feet wide. An edge of this box just touches the ladder. How far up the wall does the top of the ladder reach? (*Solution follows.*)

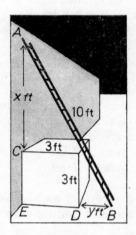

FIG. 4.1

There are different ways in which this problem may be solved; one method depends on trigonometry, but the solution below depends on algebra and geometry. Whichever method is used, however, a problem arises when it comes to solving the resulting equation. Only by taking the appropriate course at this point does it become possible to obtain a solution.

Let the top of the ladder be x feet above the box (i.e. AC in Fig. 4.1), and the foot of the ladder be y feet from the box (i.e. DB). The problem now looks quite simple. Applying Pythagoras' Theorem (the theorem is explained later in this chapter for those unfamiliar with it) to the large triangle we obtain

$$(x+3)^2 + (y+3)^2 = 100 \tag{1}$$

Also from the two similar triangles, sides which are in proportion give

$$x/3 = 3/y$$

or

$$xy = 9 \tag{2}$$

We now have two equations in two unknowns, which should be sufficient for a solution. The usual method in such cases is to eliminate one of the unknowns, and the best way here is to express equation (2) in the form

$$y = 9/x$$

We then use this to replace y in equation (1), giving

$$x^2 + 6x + 9 + \frac{81}{x^2} + \frac{54}{x} + 9 = 0$$

This simplifies to

$$x^4 + 6x^3 + 18x^2 + 54x + 81 = 0$$

Unfortunately we have arrived at a dead end as there is no easy method of solving a bi-quadratic equation such as this. However, a slightly different method of substitution produces a quadratic equation in $(x+y)$ which can first be solved, and this solution may then be used to obtain a second quadratic equation which gives the final solution of the problem. We proceed as follows:

From equation (1)

$$x^2 + 6x + 9 + y^2 + 6y + 9 = 100$$

Since, from equation (2), $2xy - 18 = 0$, we may write

$$x^2 + y^2 + (2xy - 18) + 9 + 6x + 6y + 9 = 100$$

or

$$x^2 + 2xy + y^2 + 6x + 6y = 100$$

i.e.

$$(x+y)^2 + 6(x+y) - 100 = 0$$

Putting $z = x + y$, we obtain

$$z^2 + 6z - 100 = 0$$

This may be solved by the usual formula method* to give

$$z = -3 \pm \sqrt{109}$$

Since z (i.e. $x+y$) cannot be negative, we ignore the negative value for $\sqrt{109}$, and have the two equations

$$x+y = -3 + \sqrt{109}$$
$$xy = 9 \qquad (2)$$

We then eliminate one of the unknowns from these two equations, thereby getting a further quadratic equation, and this gives the required values. The solution obtained is that the ladder reaches up the wall for a distance of 8·92 feet or 4·52 feet.

The following problem contains another interesting method of solving equations. It requires a knowledge of the elementary theory of equations which may be obtained from any algebra text-book of about sixth form level. The theory we need is given in the solution below, and a short proof will be found in the 'Hints' in Chapter 14. (*see p.* 168.)

Spot the Numbers   Find three numbers whose sum is 19, whose product is 144, and the sum of the squares of which come to 149. (*Solution follows.*)

Denoting the three numbers by a, b and c, we have

$$a+b+c = 19$$
$$a^2 + b^2 + c^2 = 149$$
$$abc = 144$$

(Notice the symmetrical nature of the relationships in each of these equations, which means that when we have obtained the three solutions, they may be applied in any order to the letters a, b and c.)   Now

$$(a+b+c)^2 = a^2 + b^2 + c^2 + 2ab + 2bc + 2ac$$

So, using the values we have been given,

$$361 = 149 + 2(ab + bc + ac)$$

whence $\qquad\qquad ab + bc + ac = 106$

---

* The solution of an equation $ax^2 + bx + c = 0$, is given by the formula
$$x = \frac{-b \pm \sqrt{(b^2 - 4ac)}}{2a}.$$
In the case of the equation above, $a = 1$, $b = 6$ and $c = -100$. Putting these values in the formula gives the solution.

And now comes our 'Theory of Equations'. If we have an equation $x^3 + px^2 + qx + r = 0$, which has roots x_1, x_2, and x_3, then:

the sum of the roots taken one at a time $(x_1 + x_2 + x_3)$ is $-p$;

the sum of the roots taken two at a time $(x_1 x_2 + x_2 x_3 + x_3 x_1)$ is q;

the sum of the roots taken three at a time $(x_1 x_2 x_3)$ is $-r$.

Applying this to our case, if a, b, c are the roots of an equation, then:

the sum of the roots taken one at a time is 19;

the sum of the roots taken two at a time is 106;

the sum of the roots taken three at a time is 144.

Hence the equation of which a, b and c are the roots is

$$x^3 - 19x^2 + 106x - 144 = 0$$

This may now be factorized as

$$(x-2)(x^2 - 17x + 72) = 0$$
$$(x-2)(x-8)(x-9) = 0$$

So $x = 2$, 8 or 9, which are the required numbers.

We sometimes meet problems which seem to offer insufficient information for a solution. Usually in such cases, some of the information which we feel we require is only of use to us in making our initial calculations, and disappears during the course of the subsequent working. By introducing a letter (x, say) to represent the missing information, we do not complicate matters, since this letter is eliminated before the final calculation. The next problem contains all the information needed for a solution.

The Narrow Bridge   In England we have many roads which we built for the leisurely progress of the horse and cart, but still use for the cars—and pedestrians—of today! A man is two-thirds of the way across a bridge which is only just the width of a car, when he sees a car approaching at 30 m.p.h. He can just manage to get safely off the bridge by running (at the same uniform speed) in either direction. What is the speed at which he must run? (*Hint p. 168. Answer p. 189.*)

Generally, when solving equations, we expect that where there are three unknown quantities, we must have three equations in order to obtain a solution. But what about this problem?

Three Unknowns Find three digits such that the product of one pair is eight and the product of the other pair is nine. (*Solution follows.*)

In the usual manner, we denote the three numbers by the letters *a*, *b* and *c*. The problem then gives the two equations

$$ab = 9$$
$$bc = 8$$

which are impossible to solve by simple substitution. There is, however, one further condition which so far has not been used: the solutions are digits, i.e. they must be whole numbers. Hence the only possible values for *a* and *b* are 1 and 9 or 3 and 3. Similarly the possible values for *b* and *c* are 1 and 8 or 2 and 4, and the only possible solution for the whole problem is 1, 8 and 9.

In this example, if we are not told that the solutions must be digits, there are a large number of possible solutions of the two given equations. A similar case is the equation $x + y = 5$. Here, even if one restricts solutions to whole numbers, there is not a unique solution. Equations such as these, where insufficient information is given to obtain a unique solution (or set of solutions), are often called diophantine equations.

Methods of solution of diophantine equations vary, but the general principle is to reduce the equations to some form where a simple substitution of successive integers will generate solutions for the original unknowns. The following two examples—both well-known—will illustrate the approach.

The first concerns numbers known as Pythagorean triples. These are sets of three numbers, where the members of each set are integral solutions of the equation $x^2 + y^2 = z^2$. Pythagoras' Theorem states that in any right-angled triangle, the square of the side opposite the right angle is equal to the sum of the squares of the other two sides. A common example is the triangle shown in Fig. 4.2 where the hypotenuse (the side opposite the right angle) is five units, and the other two sides three and four units respectively. $3^2 + 4^2 = 5^2$.

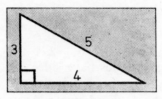

FIG. 4.2

Returning to the earlier equation, if z represents the length of the hypotenuse, and x and y the lengths of the other two sides of the right-angled triangle, the example expresses the relationship given in Pythagoras' Theorem. Hence numbers which satisfy the equation $x^2 + y^2 = z^2$ are called Pythagorean triples.

Many text books contain a proof of the method of finding values to satisfy this equation, so one will not be given here. Suffice it to say that the required triples can be generated by giving values to a and b such that a is greater than b, a and b have no common factor, and one of a and b is odd and the other even, where a and b give

$$x = a^2 - b^2$$
$$y = 2ab$$
$$z = a^2 + b^2$$

As an example, putting $a = 2$ and $b = 1$ in the above formulae, we obtain $x = 3$, $y = 4$ and $z = 5$, i.e. the 3, 4, 5 triangle mentioned above. Try other values of a and b, and check the relationship of the resulting triples.

Right-angled Triangles 1 Find a right-angled triangle whose perimeter is equal in number to its area. (*Solution follows.*)

Since the area is $\frac{1}{2}xy$ and the perimeter $x + y + z$,

$$\frac{1}{2}xy = x + y + z$$

Using the relationships above

D

$$\tfrac{1}{2}(a^2 - b^2).2ab = a^2 - b^2 + 2ab + a^2 + b^2$$

or

$$ab(a^2 - b^2) = 2ab + 2a^2$$

i.e.

$$b(a - b) = 2$$

Hence the possibilities are $b = 2$, $a = 3$ or $b = 1$, $a = 3$. The corresponding triangles have sides 5, 12, 13 and 6, 8, 10.

Right-angled Triangles 2   Find two right-angled triangles with integral length sides, which have the same hypotenuse. (*Hint* p. 169. *Answer p.* 189.)

Another familiar problem leading to a diophantine equation is the monkey and dates problem, of which there are various versions differing mainly in the number of persons or the final share-out of the dates. The general principle applies in all cases, however, so only one version is given. (You may care to make slight variations in the problem and obtain the corresponding solutions.)

The Arabs, the Monkey and the Dates   Four Arabs went on a journey across a desert and took with them a monkey and a bag of dates. On the first evening, they pitched their tent, hung the bag of dates on the tent pole and went to sleep. During the night one of the Arabs awoke and, feeling rather hungry, decided he would like to eat his share of the dates. He took the bag and counted the dates into four equal piles on the floor, but found there was one date over which he gave to the monkey. He then ate all the dates in one of the four equal piles, put the dates from the other three piles back into the bag, hung it on the tent pole and went off to sleep. The next morning, however, he forgot to mention this to his companions. They struck camp, went on across the desert and pitched their tent the following evening, once more hanging the bag of dates on the tent pole. This time, however, during the night another of the Arabs awoke and, feeling hungry, decided he would like to eat his share of the dates. He took the bag and counted the dates into four equal piles on the floor (since he was unaware of the previous share-out), found there was one date over, which he gave to the monkey, ate the dates in one of the four piles, put the rest back into the bag, which he rehung on the tent pole, and went back to sleep. The following morning he, too, forgot to mention to his companions what had happened. The story continues that the same thing happened on the two following nights with the two remaining Arabs. Then on the next day, they reached the end of their journey, and decided to share out the dates. But each had a guilty conscience and decided to say nothing of what had happened for fear of losing face. So the dates were shared out into four equal piles, there was one date over, which was given to the monkey, and each Arab ate his share of the dates. Find the least number of dates with which the Arabs could have started their journey. (*Solution follows.*)

The first part of the problem—finding an equation—is reasonably straightforward. If *n* is the number of dates at the start, the number after the first Arab's share-out is $\frac{3}{4}(n-1)$. After the second share-out, the number remaining is

$$\tfrac{3}{4}\{\tfrac{3}{4}(n-1)-1\}$$

after the third share-out it is

$$\tfrac{3}{4}[\tfrac{3}{4}\{\tfrac{3}{4}(n-1)-1\}-1]$$

after the fourth share-out it is

$$\tfrac{3}{4}\llbracket\tfrac{3}{4}[\tfrac{3}{4}\{\tfrac{3}{4}(n-1)-1\}-1]-1\rrbracket$$

and this quantity, less 1, is divisible by four. In other words

$$\tfrac{3}{4}\llbracket\tfrac{3}{4}[\tfrac{3}{4}\{\tfrac{3}{4}(n-1)-1\}-1]-1\rrbracket -1 = 4x$$

where *x* is an integer. This equation may be simplified to

$$81n-781 = 1024x$$

where both *n* and *x* are integral.

The second part of the problem is to find a solution to this equation. Re-write the equation as

$$n = \frac{1024x+781}{81}$$

$$= 12x-9+\frac{52x+52}{81}$$

For an integral value of n, $(52x + 52)/81$ must also be integral. Put

$$y = \frac{52x + 52}{81}$$

Then

$$52x + 52 = 81y$$

$$x = \frac{81}{52}y - 1$$

For x to be integral, y must be a multiple of 52. Since the question asks for the least number of dates, take $y = 52$, its lowest possible value. We then find that $x = 80$, and $n = 1021$. Hence the smallest number of dates with which the Arabs could have started is 1021.

Baling Out A farm hand is in the corner of a field with a truck fastened to the back of his tractor. He has loaded a number of bales of hay on the truck, but finds he is facing the wrong way to get out of the gate to the field. He decides, therefore, to drive round the perimeter of the field until he arrives back at the gate, but being a reckless driver, when he comes to the first corner of the field, he swings his tractor round so quickly that some bales fall off. The number falling off is one-third of his load and one-third of a bale. He again takes the following corner too quickly, and now a third of his remaining load and a third of a bale fall off. In all, this happens four times before he gets back to the gate: at each corner a number of bales fall off, which amount to a third of the bales that were on his truck as he came up to the corner, and an extra third of a bale falls off. Yet all the bales remained whole, none of them being split in falling, and subsequently he was able to re-load them by lifting them onto his truck. What is the least number of bales with which he started, and how many bales remained when he stopped at the gate? (*Hint p.* 169. *Answer* p. 189.)

Sum and Product Find all the two digit numbers which are multiples of the product of their digits. (*Hint p.* 169. *Answer p.* 189.)

Small Change In how many different ways can I make up the sum of 80p. using only two-pence and five-pence pieces? (*Hint p.* 170. *Answer p.* 189.)

At times we meet problems where the more usual equations are replaced by inequalities. As a simple example, we might wish to find two numbers whose sum is less than fifteen and, denoting the numbers by a and b, we write $a+b<15$. Note that this does not include the sum which is fifteen; in such cases we say 'is equal to or less than', e.g. two numbers whose product is equal to or less than thirty-six, is denoted by $ab\leqslant 36$. We may have a number of such conditions applying to any one particular problem, and we find that various solutions may be found which fall within the desired limitations. The problem then is to find the maximum (or minimum) values from the various possibilities.

In recent years this type of problem has become common in industry, commerce, transportation, etc., and can be of a complex type which frequently requires solution by computer. However, simple problems and puzzles may be solved by calculation or by graph, as in the following example. The method employed in these problems is known as linear programming.

Washday Blues   A manufacturer makes two kinds of detergent, Swoosh and Roto. One hundredweight of Swoosh goes through the mixer in fifteen minutes and then takes ten minutes in the dehydrator. One hundredweight of Roto, on the other hand, only takes twelve minutes in the mixer, but fifteen minutes in the dehydrator. In all he has six mixers and five dehydrators each working an eight-hour day. If his profit on Swoosh is ninety new pence per hundredweight, and his profit on Roto is one pound per hundredweight, how should he arrange his production so that he obtains maximum profit? (*Solution follows.*)

A graphical method of solution will be given. First the relationships between the variables are obtained.

Let s hundredweights be the daily production of Swoosh and r hundredweights the daily production of Roto.

Then the total time taken in the mixer by these quantities will be $\frac{1}{4}s+\frac{1}{5}r$ (in hours). But since there are six mixers available for the eight-hour day, the total time available is forty-eight hours. Our expression, therefore, may have any value up to forty-eight hours, but cannot exceed this amount. This gives

$$\tfrac{1}{4}s + \tfrac{1}{3}r \leqslant 48$$

Proceeding in a similar fashion for the dehydrator, we obtain

$$\tfrac{1}{6}s + \tfrac{1}{4}r \leqslant 40$$

These two equations may be simplified to

$$5s + 4r \leqslant 960 \tag{1}$$
$$2s + 3r \leqslant 480 \tag{2}$$

Certain other conditions are imposed by the problem. Thus r and s cannot be less than zero. So

$$r \geqslant 0 \tag{3}$$
$$s \geqslant 0 \tag{4}$$

Finally, the total profit is given by the equation

$$90s + 100r = P \tag{5}$$

where P is in new pence.

Now draw the graphs. The inequality (1) is represented by the area below the line $5s + 4r = 960$ (Line 1 in Fig. 4.3), and (2) by the area below the line $2s + 3r = 480$ (Line 2 in the diagram). The inequalities (3) and (4) mean we need not consider negative values of r and s.

From what has been said, our solution must lie within the shaded area. Taking any convenient point on the r-axis, say $r = 90$, and substituting this value in (5), we get $P = 9000$ (since $s = 0$ on the r-axis). Putting $r = 0$ and keeping $P = 9000$, we get the point $s = 100$ on the s-axis. Joining these two points gives us a line representing equation (5). Any point on this line will give values for r and s corresponding to a profit of 9000 new pence; we call the line the 9000 p. profit-line.

If we give P other values we obtain a series of profit lines, each one of which is parallel to the one we have just drawn. The further the profit line is from the origin of the graph (i.e. where r and s both equal zero), the greater is the value of P, or the greater the profit. The problem asks for the maximum profit, so we must draw a line parallel to the 9000 p. profit-line in such a way that it still falls within our shaded area (or on its boundary) and is as far as possible from the origin. This line

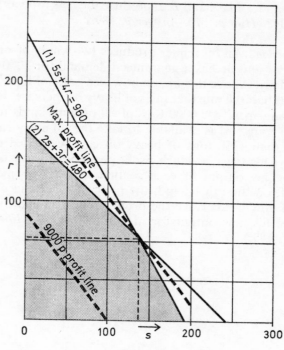

FIG. 4.3

gives the quantities for the maximum profit. As can be seen, it passes through the point of intersection of Lines 1 and 2, and (to the nearest 5 hundredweights) the answer to the problem is that for maximum profit, the manufacturer should make 70 hundredweights of Roto, and 135 hundredweights of Swoosh.

Food Mixer An animal foodstuff manufacturer intends to market a pack of a new food which will contain sufficient quantities of two nutrition elements N_1 and N_2 to satisfy an animal's daily requirements, estimated at 120 units of N_1 and 68 units of N_2. He is going to make the new food from two existing foods *A* and *B*, whose content per ounce of N_1 and N_2 and cost are shown below.

	Units of N_1 per oz.	Units of N_2 per oz.	Cost per oz.
Food *A*	12	4	1p.
Food *B*	5	4	$\frac{1}{2}$p.

How should he mix *A* and *B* to produce the new food as cheaply as possible? (*Hint p.* 170. *Answer p.* 189.)

Oil Production   An oil refiner produces two kinds of oil—light and heavy—and he has a guaranteed demand for 3000 tons of light oil daily. Above this figure, however, sales of light oil only increase at half the rate that sales of heavy oil increase. His daily production capacity is 6000 tons of oil of both kinds together. Although transport is available to take the oil away each day, no more than 4250 tons of heavy oil can be carried daily. If light oil yields twice as much profit per ton as heavy oil, what should be his output of each so that he achieves maximum profit? If, on the other hand, the price of oil changes so that heavy oil now yields twice the profit per ton as light oil, what should be his new output for maximum profit? (*Hint p.* 170. *Answer p.* 189.)

CHAPTER 5

Interlude 1: Something to Think About

First A monkey is holding on to a rope which passes round a frictionless pulley and supports at the other end a weight which is just as heavy as the monkey. The monkey now starts to climb the rope. What happens? (*Answer p.* 189.)

Second A boat is afloat on a lake, and in the bottom of the boat is a quantity of scrap iron. The scrap iron is now thrown overboard into the lake. Does the level of the lake rise, fall, or stay as it is? (*Answer p.* 190.)

Third Fig. 5.1 shows a pulley over which hangs a string supporting at each end two equal pulleys. Each of the latter pulleys

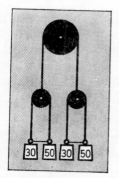

FIG. 5.1

has a string over it, supporting at one end a weight of 30 grammes and at the other end a weight of 50 grammes. Initially the two weights supported by one pulley are looped together with a piece of thread, and similarly the two weights around the other pulley. Clearly, at this moment, the whole system is in balance. One of the supporting loops is now cut, thereby releasing the two weights that it was holding stationary. Describe the motion. (*Answer p.* 190.)

Fourth A firm in north-eastern England makes very large transformers and sends them by road to its customers. The firm has recently bought a special transporter which works on the hovercraft principle, lifting the load on a cushion of air. The idea is that by using this transporter it will not be necessary to spend money in strengthening bridges over which the transformers have to pass on their journeys. But there is a fundamental principle of mechanics that forces have to balance; in other words, if the air under the hovercraft is pressing up on the load, it must be pressing down with equal force on the ground beneath. Is the firm's idea a sound one? (*Answer p.* 190.)

Fifth A wooden cube, whose edges are three inches long, is to be cut into twenty-seven 1in. cubes. If, after each cut with a saw, the resulting pieces may be rearranged in any desired way, what is the smallest number of saw-cuts needed to produce the desired result? (*Answer p.* 190.)

Sixth Test the truth of the following:

Statement 1: All rules have exceptions.
Statement 2: Statement 1 is a rule.
Statement 3: It follows that Statement 1 has exceptions.
Statement 4: Hence, all rules do not have exceptions.

(*Answer p.* 190.)

CHAPTER 6

Set Fair

Motorists' Choice   The manufacturers of Zoom petrol and oil carried out a survey recently by stopping a number of motorists at random and asking them what brand of petrol and oil they used. One-third of those interviewed did not use Zoom petrol, and three-eighths of the total interviewed did not use Zoom oil. In fact, fifty-nine motorists used both Zoom petrol and oil, but one-fifth of the total interviewed did not use either the petrol or the oil of Zoom manufacture. How many motorists were interviewed? (*Hint p*. 171. *Answer p*. 190.)

If you asked someone with slight knowledge to say what were the topics of the newer mathematics now being introduced into schools, you would most likely get an answer which included mention of 'sets'. Sets is probably the most widely known topic of this newer work because it is frequently the first topic to be covered, it has an appeal to newcomers to the subject, and it is quite easy to understand. In any case, a knowledge of sets is fundamental to an understanding of mathematics.

Any collection of objects is a set. We may actually call it a set (e.g. a set of knives and forks, a set of chessmen), or we may have some other group name for it (e.g. a flock of sheep, a football team), or we may have no particular name for it at all (e.g. the term 'my house' may be taken to include the building and all my possessions within it). In addition, we can make up a set from any collection of objects which may have no apparent connection (e.g. this book, the Houses of Parliament, and the pretty girl who lives down the street at number 27).

Normally we denote a set by a letter, and list the parts of it (called the elements) in curly brackets (called braces). For example, A may be the set of odd numbers from one to twelve, and we could write this in various ways, thus

$$A = \{\text{All odd numbers from 1 to 12}\}$$
$$A = \{1, 3, 5, 7, 9, 11\}$$

We could now take another set B, consisting of prime numbers from one to fifteen:

$$B = \{1, 3, 5, 7, 11, 13\}$$

Comparison of A and B shows that certain numbers are contained in each set, i.e. 1, 3, 5, 7, 11, and this is called the intersection of the two sets, written $A \cap B$, thus:

$$A \cap B = \{1, 3, 5, 7, 11\}$$

On the other hand, if we make up a set consisting of all the elements contained in one or both of the two sets, we have the union of sets A and B, written $A \cup B$, thus

$$A \cup B = \{1, 3, 5, 7, 9, 11, 13\}$$

Although work on sets may be done using this notation, it is easier for the beginner, and adequate for the purpose of solving the problems in this book, if we use a system of diagrams to represent sets, their intersection and their union. Such diagrams are called Venn diagrams, and consist of a number of overlapping closed curves (often circles). Each of these closed curves can represent a set. Thus in Fig. 6.1 we have two sets, A and B, which overlap. The central part, which lies in both set A and set B, represents $A \cap B$, while the union of the sets, $A \cup B$, consists of the total area lying within the outside boundary curve. Taking sets A and B to be defined as above, insert the numbers 1, 3, 5, 7, 9, 11, 13 in the appropriate spaces in the diagram.

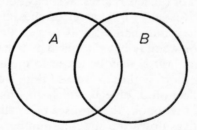

Fig. 6.1

It will be found that the majority of the numbers (1, 3, 5, 7 and 11) lie in the central part, only the number 9 lying in the part of A outside B and only 13 lying in the part of B outside A.

If there are three sets A, B and C, provision has to be made for all possible intersections. These are the intersections $A \cap B$,

$B \cap C$, $C \cap A$, and also the intersection of all three $A \cap B \cap C$. This is usually achieved by a diagram as shown in Fig. 6.2 (you should identify the various spaces).

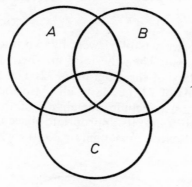

FIG. 6.2

The representation of more than three sets on a Venn diagram will be considered in greater detail in a later chapter, as it has certain interesting geometrical aspects. But as an example, check whether Fig. 6.3 is adequate for the representation of four intersecting sets.

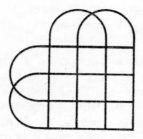

FIG. 6.3

Various problems follow which use Venn diagrams in their solution, but it is possible to solve many of these by other methods. In general, however, the use of sets and Venn diagrams results in a neater and easier method of solution.

The TV Set A survey was made of one night's viewing by seventy-two television viewers, who were asked to state whether they had watched BBC and ITV programmes that evening. Two-thirds replied that they had watched ITV at some time during the night, and one-eighth had watched both BBC and ITV. How many had watched only the BBC programmes? (*Solution follows.*)

Represent the two sets of viewers by two intersecting circles, as shown in Fig. 6.4. Then the total number of viewers within the outer boundary of the diagram is 72. The ITV circle must include two-thirds of 72, i.e. 48 in all. Also the intersection of the two circles must include one-eighth of the total viewers, i.e. 9. Putting this 9 into its appropriate space first, we see that the rest of the ITV circle must contain 39 (the whole circle has to contain 48).

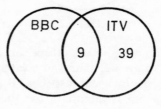

FIG. 6.4

Now fill in the rest of the diagram. The answer is that 24 viewers saw BBC programmes only.

A Further TV Set On another evening sixty-five families were asked to give details of the programmes they had viewed, and list them as from BBC-1, BBC-2 and ITV. Twenty-eight said they had watched some BBC-1 programmes, forty-one saw some ITV programmes, twenty saw BBC-2, and one family never viewed at all that evening. Of course, there was some overlap of viewing since some families watched programmes from more than one source. Ten said they had seen both BBC-1 and BBC-2 programmes, twelve said they had seen both BBC-1 and ITV, and a further twelve had seen some programmes from ITV and some from BBC-2. How many had watched all three programmes, ITV, BBC-1 and BBC-2? (*Solution follows.*)

In this problem there are three sets, BBC-1, BBC-2 and ITV,

so we need three intersecting curves to represent each of these
and the various intersections of sets, as shown in Fig. 6.5. Let
x be the number of families who saw all three programmes, and
insert this in the appropriate space on the diagram.

Now fill in the other intersections. 10 saw BBC-1 and BBC-2,
and this must include x who also saw ITV; so the number who
saw BBC-1 and BBC-2 but not ITV must be $10-x$. Similarly
the number viewing BBC-1 and ITV, but not BBC-2, must be
$12-x$, and those who saw ITV and BBC-2, but not BBC-1,
must be $12-x$.

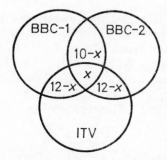

FIG. 6.5

Consider now the 28 who viewed BBC-1. The total of the
numbers lying in the circle representing BBC-1 must be 28.
At present there are $(10-x)+x+(12-x)$ already in this circle,
i.e. $22-x$. It follows that in the part of the circle still blank we
must insert $6+x$ in order to make up the total of 28, since
adding $22-x$ to $6+x$ gives 28. Proceeding similarly with the
other two spaces for BBC-2 and ITV, the final diagram is
shown in Fig. 6.6.

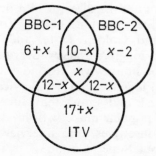

FIG. 6.6

ow that the total number of viewing families was
ily did not view at all), the sum of the numbers in
parts of the circles must be 64. This gives.

$$-x)+(x-2)+x+(12-x)+(12-x)+(17+x) = 64$$

$$55+x = 64 \quad \text{or} \quad x = 9$$

so the number of families who viewed all the programmes was
nine.

Light Reading   Of 100 readers of fiction borrowing books from
a library, 38 liked historical novels, 58 liked thrillers, and these
included 6 who liked both. Moreover, 12 liked both these and
adventure stories also. Of those who liked historical novels, 4
also liked adventure stories. In all 52 readers liked adventure
stories. How many liked both thrillers and adventure stories,
and how many liked adventure stories only? (*Hint p*. 172.
Answer p. 190.)

The College Students   In a certain college, 60 per cent of the
students live in college, 75 per cent of the students are female,
and 68 per cent of the total college population are Arts students.
Five per cent of the total students are female Arts students who
live in college, and there are ten times as many female students
who live in college and do not study Arts, as there are male
students who live in college. Find what percentage of the college
population are female Arts students who do not live in College.
(*Hint p*. 172. *Answer p*. 190.)

Market Survey   You are the director of a firm of butter manufacturers who produce Sunshine butter. Your main rivals are
Super Goodness and Golden Taste, and you have recently
employed Messrs Floggitt's Survey Services to do a survey
among a sample of housewives to find the brands of butter
which they buy. Floggitt's Survey Services report has just
arrived on your desk and shows that Sunshine butter is well
ahead as the most popular make with housewives. The report
states that 1500 housewives were interviewed, and 390 never
bought anything but Super Goodness, while 203 stated that
they only bought Golden Taste. On the other hand 460 house-

wives were so satisfied with Sunshine butter that they never bought anything else. Ninety-three housewives said they bought all three makes from time to time, and 165 sometimes bought Super Goodness and Golden Taste, but never Sunshine. But 72 housewives who bought Super Goodness and 81 housewives who bought Golden Taste also often bought Sunshine. Finally, the total number of housewives who at some time or other bought Super Goodness was 620, the total number who bought Golden Taste at some time or other was only 542, while the number of housewives who stated that they bought Sunshine butter, either regularly or from time to time, was no less than 780. 'Our survey' write Floggitt's Survey Services, in conclusion, 'has confirmed the popularity and supremacy of your product with the housewives we interviewed.' What do you reply? (*Hint p.* 173. *Answer p.* 190.)

Language Laboratory In the Arts sixth form of our school there are sixty-five pupils who are studying at least one foreign language. Twelve are studying both French and Latin, but no other language, two are taking Latin and German only, and ten Latin and Spanish only. Of the others taking only two languages, seven are taking French and German, six are taking French and Spanish, and one only is studying German and Spanish. There are no pupils studying Latin, German and Spanish, and none taking Latin, German and French. The number of pupils studying French, German and Spanish, but not Latin, is the same as the number in each case who are studying one subject only, Latin, French, German or Spanish. Six pupils are taking French, German and Spanish, but this number includes some who are taking Latin as well, and seven pupils are taking Latin, French and Spanish, although again this number includes those taking all four subjects. What is the total number of pupils who are taking one language only? (*Hint p.* 173. *Answer p.* 190.)

Some problems on logic may be solved by the use of Venn diagrams. A later chapter will deal particularly with the solution of logical problems, so at this stage we shall consider only some simple examples which may be solved by the use of sets.

Before proceeding, it is necessary to explain some further ideas. If one set, *A*, is contained entirely within another set,

E

B, then A is said to be a sub-set of B, and this is written $A \subset B$; the corresponding Venn diagram is shown in Fig. 6.7.

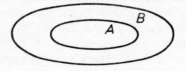

FIG. 6.7

At times we need to define our 'universe', i.e. the particular things about which we are talking. Thus if we consider the set of boys, R, in a certain class at school who have red hair, and another set of boys, S, in the same class who wear spectacles, then our universe would likely be the particular class of boys, and we show this on a Venn diagram by drawing a rectangle around our other sets and calling it U (Fig. 6.8).

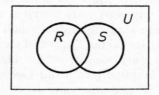

FIG. 6.8

The use of the universe is particularly important when we consider the inverse of any set, e.g. boys in the class who have not got red hair (we usually denote this by adding a mark ' after the normal sign; i.e. R denotes boys with red hair, and R' denotes boys who have not got red hair). In Fig. 6.9 the shaded area represents boys in the class who have not got red hair.

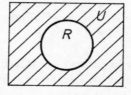

FIG. 6.9

Further applications will be found in the problems which follow.

Example Represent on a Venn diagram the statement 'All judges are wise men'. (*Solution follows.*)

There are two sets involved in this statement: the set of wise men and the set of judges. Since the statement says that *all* judges are wise men, the set of judges must be contained within the set of wise men, i.e. the set of judges is a sub-set of the set of wise men. The diagram is shown in Fig. 6.10.

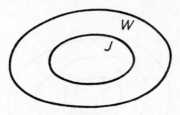

FIG. 6.10

What is represented by the space in the diagram which lies outside the set J of judges but within the set W of wise men? Notice from the diagram that the converse of the statement (i.e. that all wise men are judges) is not necessarily true. Nor does it follow that only judges are wise men.

Drawing false conclusions, such as those disproved in the last sentence, leads to a great deal of fallacious argument in everyday life, and is far more common than we realize. The logic of sets illustrates such mistakes; it also shows the need for precise statements, as we shall see in the second example below.

Through Trains

1. All trains from London to Manchester are electric.
2. All trains from London to Manchester pass through Crewe.

Hence all trains which pass through Crewe are electric.

Use a Venn diagram to test the truth or falsity of that conclusion.

(*Hint p.* 173. *Answer p.* 190.)

The Dog Tree

   1.   All things which have a bark are dogs.
   2.   These trees have a bark.

Hence these trees are dogs.

Test the truth of that assertion. (*Solution follows.*)

Let *B* be the set of things which have a bark, and *T* the set of these trees. If *D* is the set of dogs, we then have the diagram shown in Fig. 6.11.

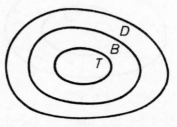

FIG. 6.11

This shows that our conclusion is true.

We know, of course, that this is a play on words and that the conclusion is false; our common sense tells us that. The fault lies in our initial statements and not in the logic and the working of the problem. Given Statements 1 and 2 and no others, then our conclusion is correct.

School Subjects

Draw a conclusion from the following statements about a certain class at school:

   1.   All pupils do history, or geography or both.
   2.   There are no pupils who do both history and mathematics.
   3.   Every pupil taking English also takes geography.

(Solution follows.)

Let *H* be the set of pupils taking history, *G* those taking geography, *E* those taking English and *M* the set taking mathematics; then the Venn diagram is as shown in Fig. 6.12. (Notice that *M* lies entirely within *G* since *all* pupils are included within the two sets *G* and *H*—statement 1.)

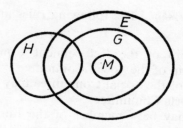

FIG. 6.12

The conclusion is that every pupil taking mathematics also takes English.

Top People

Top People read *The Times*.
The Times costs 5p.
No working man can afford 5p. for a daily paper.

Test the truth of the conclusion that working-class men are not top people. (*Hint p.* 173. *Answer p.* 190.)

Office Routine

1. It is a rule of this office that no girl under 18 may be a typist.
2. All the hard-working girls in this office are brunettes.
3. None of the girls who are not typists can do shorthand.
4. In this office, none but those under 18 are slackers.

Draw a conclusion from the above statements. (*Hint p.* 174. *Answer p.* 190.)

Animal Crackers

1. Unicorns are legendary animals.
2. There are many absurd animals, including legendary ones.
3. Giraffes are absurd.
4. Giraffes are the only absurd animals that I like.

Test the truth of the conclusion that I don't like unicorns. (*Hint p.* 174. *Answer p.* 191.)

Committee Membership The following rules apply at the local club:

1. Membership of the Refreshment Committee includes membership of the Social Committee.
2. Members of the Social Committee may not be members of the Sports Committee.
3. Anyone may be a member of the Finance Committee provided that he is not a member of the Refreshment Committee.
4. Members of the General Committee are members of both the Social and Finance Committees.

Draw a conclusion from the above rules. (*Hint p.* 175. *Answer p.* 191.)

CHAPTER 7

Which Way? And How Far?

Flight Path An aircraft is at a certain point at a given latitude on the Greenwich meridian. The captain wishes to fly the aircraft to a point with the same latitude on the 180° meridian, and notices that it would be exactly the same distance if he flew over the North Pole, as if he flew along the line of latitude. What is the latitude of the point?

Higher Communications A telegraph wire is laid around the surface of the earth at the equator, and a similar wire is laid on the moon's surface around its equator. (For the purpose of this problem, both the earth and the moon may be regarded as perfect spheres.) It is now required to raise each of these wires six feet above the surface of the earth and moon respectively, supporting each on poles (we assume there is no sag between the poles). Clearly, in each case more wire is required in order to do this, but which needs the greater amount of extra wire, the earth or the moon?

Tangential Chord Two concentric circles are such that a ten-inch chord of the outer circle is a tangent to the inner circle.

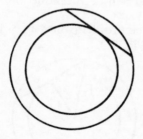

FIG. 7.1

What is the area between the two circles?

(*Solutions follow.*)

These three problems have something in common, insofar as they appear to contain inadequate information for their solution. Should we not be given the radius of the earth in the first example, the radius of both the moon and the earth in the second, and the radius of the inner (or outer) of the two concentric circles in the third? Yet further reference is unnecessary; all the required information is given in the question in each case.

The first problem does not require any calculation. It is a well-known fact of spherical geometry that the shortest surface distance between any two points which lie on a sphere is along the great circle between the two points. A great circle is the largest circle which can be drawn on a sphere; its diameter is the same as the diameter of the sphere. Although every meridian is a great circle, the equator is the only circle of latitude which is a great circle. So if it is just as far for the aircraft to go via the North Pole as it is to fly along the line of latitude, then—since the North Pole route is a great circle—the route along the line of latitude must be a great circle. It follows that the aircraft must be on the equator.

To solve the second problem, consider a sphere of radius R feet with the wire along its surface. The length of the wire will be $2\pi R$ feet. When the wire is raised a distance of 6 feet above the surface, the new radius is $(R+6)$ feet, and the length of the circular wire will be $2\pi(R+6)$ feet, which equals $2\pi R + 12\pi$ feet. So the difference in length is 12π feet, which is independent of R. Hence, in each case—on the earth and on the moon—the extra length of wire needed will be the same and will be 12π feet.

The last of the three problems may be solved by drawing the two radii shown in Fig. 7.2.

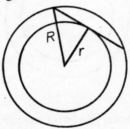

FIG. 7.2

Let the radius of the outer circle be R, and the radius of the smaller circle be r. Applying Pythagoras' Theorem to the right-angled triangle in the figure,

$$R^2 - r^2 = 5^2 = 25$$

But the difference in area of the two circles is

$$\pi R^2 - \pi r^2 \quad \text{or} \quad \pi(R^2 - r^2)$$

which from the result above equals 25π.

However, a much easier method of solution of this last problem is based on logical argument rather than mathematical calculation. Since the problem has apparent lack of information (we are only given the length of the chord) it would seem likely that the ultimate result is independent of the size of the circles. Suppose we imagine them decreasing in size, with the chord remaining ten inches long, until the inner circle becomes a point circle, i.e. its radius becomes zero and the circle is simply the point which is the centre of the larger circle.

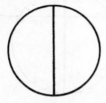

Fig. 7.3

The chord has now become the diameter of the larger circle (Fig. 7.3), and the area of this circle is now 25π. This, therefore, must be the required area between the two circles whatever size they are, provided that the chord is ten inches long.

The Floating Ball A ten-inch diameter ball is floating, partly submerged, in a pond. If the top of the ball is two inches above the surface of the pond, what is the length of the circumference of the circle where the surface of the water is in contact with the ball? (*Solution follows.*)

The key to this problem lies in a simple geometric property proved in most school geometry books: if two chords of a circle cut each other, so that one chord is divided into two parts of lengths a and b, and the other chord is divided into two parts of lengths x and y, then $ab = xy$.

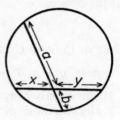

FIG. 7.4

In the present case we take a vertical section through the ball. The surface of the water may be represented by a horizontal line, AB in Fig. 7.5, and AB will be the diameter of the circle formed by the surface of the water on the ball. This is the circle whose circumference we require.

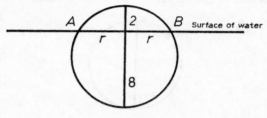

FIG. 7.5

Call r the radius of this circle. Then $r = \frac{1}{2}AB$. We now have two chords of the circle perpendicular to each other: the vertical diameter of the sphere, and the horizontal diameter, AB, of the circle of contact. Since the top of the ball is two inches above the water level, the measurements are as shown in the diagram and, using the property of chords given above,

$$r^2 = 2 \times 8 = 16$$

Hence r is four inches, which gives the required circumference as 8π inches.

Navigable Waterways Anthony was recently bitten by the sailing bug, and built himself a boat. He lives a short distance inland, but fortunately a river runs past his house and he intended to use this to sail his boat down to the sea. Less fortunately, between his house and the sea there is a bridge over the river, the arch of which forms an exact semi-circle of diameter twenty feet with the water surface. The height of Anthony's mast above the water-level was 9½ feet. This was all right, thought Anthony, until he found, to his dismay, that there was a tow-path 8½ feet wide on one side under the bridge. Now the beam of his boat was 6 feet. Could he safely pass under the bridge? (*Hint p. 175. Answer p. 191.*)

Traditional school geometry is concerned with finding relationships between lines, angles and various (mostly two-dimensional) figures, which we call the properties of the figures. The modern approach has a different emphasis. We are still interested in relationships, but not just relationships within a figure. We want to move the figures about, enlarge them, diminish them and change them from one shape to another; in other words, we are studying relationships in space.

Earlier in the book, we investigated tesselations (Chapter 3). There we were concerned with filling two-dimensional space, and we fitted shapes together so that we covered a surface. The following example has some relationship to this earlier work.

Triangulation Take any triangle, *T* say, and another, *t*, exactly the same size and shape (cut them out of card, if you wish). How many different shapes may be formed from these two triangles by putting equal sides together? You are allowed to turn the triangle over, i.e. reflect it. One example is given in Fig. 7.6.

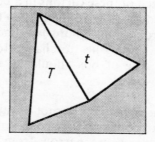

FIG. 7.6

Now find the number of different shapes which can be made in the same way using 3, 4, 5, . . . , *n* triangles. (*Hint p.* 175. *Answer p.* 191.)

Space is three-dimensional, so the study of geometry must not be restricted to plane surfaces and plane figures; it must also investigate solid objects and three-dimensional space. It is interesting to see what types of solid figures will completely fill a space. Cubes are an obvious example, but what other solids will do so? This is the parallel problem in three dimensions to tesselations in two dimensions. It has certain parallels, also, in the results. For example, we found there were certain shapes which would tesselate a surface by themselves (e.g. triangles), and in other cases we could tesselate using two shapes together (e.g. regular octagons and squares with sides the same length.) In a similar way there are certain solids (e.g. cubes, hexagonal prisms) which will fill space by themselves, and combinations of two solid figures which fit together to fill space (e.g. truncated cubes and octahedrons; triangular prisms and hexagonal prisms). The subject is rather complex, but an interesting field of investigation for the enthusiast. The work has important applications in such matters as architecture and packaging of materials.

Packaging has become a very important factor in the commercial world; vast amounts of money are spent on it annually, and the effectiveness of packaging can govern costs, demand and sales.

Packaging A manufacturer packs his product in rectangular-faced cartons measuring 2 inches by 8 inches by 10 inches. He then packs four dozen of these cartons in a box with rectangular faces. What size should he have the box so that it uses a minimum amount of cardboard? (Ignore flaps and other overlapping parts.) Suppose that he wishes to change to boxes containing fifty cartons, what size of box should he use? (*Hint p.* 176. *Answer p.* 191.)

Design factors in packaging do not always enable the most economical shape to be used. Take soap powder, for example; tall, wide cartons which are narrow from front to back are preferred to the more economical cubic shape, yet the cube has less area and therefore uses less carboard for any given volume.

The shape which has the minimum surface area for a given volume is the sphere, so the cost of packing in spherical form would be less than for any other shape, as far as material costs are concerned. Whether it costs more to mould spheres than to make rectangular cartons is another matter. The shape with the greatest area for any given volume is the regular tetrahedron. Yet many milk suppliers are using this shape of carton; clearly other advantages must operate in their case.

Fitting shapes, one within another, is an aspect of geometry with results which are sometimes novel and unexpected. This is not the same problem as that discussed earlier where we filled spaces of indefinite size with shapes. As an example of what is meant in the present case, take any cube and draw the diagonals of its faces. Together these form two intersecting regular tetrahedra. One of these is shown in Fig. 7.7 (*ABCD*), and it

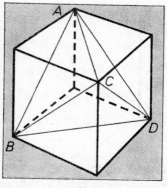

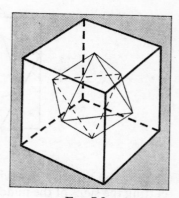

FIG. 7.7

FIG. 7.8

will be seen that this fits into the cube so that each of its edges lies on one face of the cube.

If, instead, we join the mid-points of the faces (i.e. where the diagonals cross) we obtain a regular octahedron. In this case each of the vertices of the octahedron lies on a face of the cube.

Some simple exercises follow, using the idea of shapes fitted into shapes. Part of the exercise lies in reaching quick solutions.

Double Inscription 1 A square is inscribed in a square which is inscribed in another square as shown in Fig. 7.9. If the outer square is of side two inches, what size is the inner square?

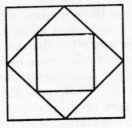

FIG. 7.9

Now suppose the middle square to be arranged so that its vertices do not cut the outer square at the mid-points of the sides, but in the ratio 1 : 2, as shown in Fig. 7.10. What is the new area of the inner square? Repeat with other values and generalize for division in the ratio 1 : n.

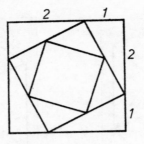

FIG. 7.10

Repeat the exercise but with the vertices of the inner square cutting the sides of the middle square in the same ratio as that in which the vertices of the middle square cut the sides of the outer square (Fig. 7.11). Generalize the result. (*Hint p.* 176. *Answer p.* 191.)

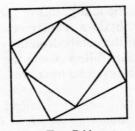

FIG. 7.11

Double Inscription 2 The problem is similar to the one above, but the intermediate square is replaced by a circle. Find the size of the inner square if the side of the outer one measures 2 inches. (*Hint p.* 177. *Answer p.* 191.)

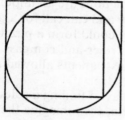

FIG. 7.12

Double Inscription 3 We again alter the diagram, and have a circle inscribed in a square, which in turn is inscribed in a circle (Fig. 7.13). If the outer circle is of diameter two inches, what is the diameter of the inner circle? (*Hint p.* 177. *Answer p.* 191.)

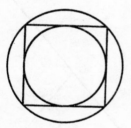

FIG. 7.13

We next develop our problems into three dimensions.

Cubism 1 A cubical box is fixed within a sphere so that each of its vertices is in contact with the sphere. The sphere, in turn, is contained in a cubical box so that it just touches each of the faces of the cube. If the outer cubical box has a side of two inches, what is the length of the side of the inner cubical box? (*Hint p.* 177. *Answer p.* 191.)

Cubism 2 A sphere is contained within a cubical box so that it touches each of the faces of the box. The cube is suspended within a second sphere so that each of its vertices touches the sphere. If the diameter of the outer sphere is two inches, what is the diameter of the inner sphere? (*Hint p.* 177. *Answer p.* 191.)

Geometrical constructions are an important part of the study of Euclidean geometry and can call for a great deal of ingenuity in their solution. They could form a puzzle topic in themselves; most are regarded as 'ruler-and-compass' constructions, which means that the only instruments allowed are compasses and an ungraduated straight edge.

Probably the best way of finding a solution is to do what part of the problem one can, then imagine the completed figure and look for relationships in it which enable us to carry on with the construction. This is illustrated by the way in which we may attempt to construct a triangle when we are given the length of a side *AB*, an angle *A*, and the perimeter. An obvious start is to draw *AB*, construct the angle *A*, and draw a line *AD* of indefinite length knowing that somewhere along it *C* must lie.

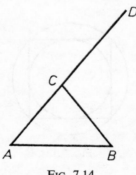

FIG. 7.14

Next we mark a point *C* on *AD* where we suppose *C* would lie if the construction had been done, then join *BC*, and look for some geometric relationship. Notice that we have not yet made use of the given length of the perimeter; as we know *AB*, this means we can find *AC* + *CB*. If we measure this along *AD*, and suppose this total length was *AE*, *CE* would equal *CB*.

Hence *ECB* is an isosceles triangle. So if we draw the perpendicular bisector of *EB*, it will pass through *C*. The construction method is now evident: draw *AB* and angle *A*, and make *AE*

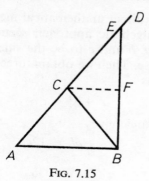

FIG. 7.15

of length equal to that of the perimeter less *AB*. Join *EB* and draw its perpendicular bisector *FC* to cut *AE* in *C*. The required triangle is *ABC*.

Drawing Exercise 1 Construct a right-angled triangle, given that the length of one of the sides (not the hypotenuse) is twelve inches, and the perimeter of the triangle is 36 inches. (*Hint p. 177. Answer p. 191.*)

Drawing Exercise 2 Construct a triangle given the perimeter, the length of the base and the altitude. (*Answer p. 192.*)

Triangle Construction In *Drawing Exercise 1*, the side of 12 inches, being adjacent to the right angle, is an altitude of the triangle. Construct a right-angled triangle with perimeter 36 inches and altitude 12 inches, but this time with the altitude from the right angle to the hypotenuse. (*Solution follows.*)

Purists would say that one should use only compasses and ruler for geometrical constructions, but nowadays we tend to

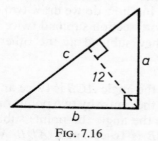

FIG. 7.16

F

agree that the most direct mathematical method is the best. In this example, an algebraic approach seems to give the best solution. Suppose a, b and c to be the sides of the triangle, c being the hypotenuse. Then we obtain three equations: Pythagoras Theorem gives

$$a^2 + b^2 = c^2 \qquad (1)$$

The area of the triangle gives

$$ab = 12c \qquad (2)$$

The perimeter gives

$$a + b + c = 36 \qquad (3)$$

By simple multiplication

$$\begin{aligned}(a+b)^2 &= a^2 + b^2 + 2ab \\ &= c^2 + 24c \quad \text{from (1) and (2)} \\ (36-c)^2 &= c^2 + 24c \quad \text{from (3)} \\ 1296 - 72c + c^2 &= c^2 + 24c \\ 96c &= 1296 \\ c &= 13\tfrac{1}{2}\end{aligned}$$

Hence

$$\begin{aligned}a + b &= 22\tfrac{1}{2} \\ ab &= 162\end{aligned}$$

Eliminating a,

$$2b^2 - 45b + 324 = 0$$

and solving for b by formula we obtain

$$b = \tfrac{1}{4}(45 \pm 9\sqrt{57})$$

The $+$ and $-$ signs give two solutions to this equation which are the values of a and b in the two possible cases.

The Intersecting Circles We often have examples of two circles cutting each other, but how do we draw two such circles so that the two points of intersection subtend twice the angle at one of the centres that they subtend at the other centre? (*Solution follows.*)

We require that the angle ACB is twice angle AOB. Since the angle subtended at the centre by two points on the circumference of a circle is twice the angle the points subtend at the circumference, angle ACB is twice angle ADB. We also know that

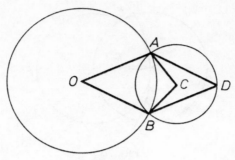

FIG. 7.17

$OA = OB$. Begin, therefore, by drawing the rhombus $AOBD$ with angle AOB the smaller of the two angles at the centres. One circle is then drawn with centre O to pass through A and B and the other circle is drawn to pass through the three points A, B and D.

Line Bisection One of the best-known geometrical constructions is that of bisecting a given line using ruler and compasses. Show how to find a point midway between two other points, using only a pair of compasses in the construction. (*Solution follows.*)

In this case we are not allowed the use of a ruler, nor are we given the straight line which we are to bisect; we are simply given two points and we have to find the point which would bisect the line joining them. This may seem a novel requirement to those familiar with Euclidean geometrical constructions, but at the turn of the seventeenth century Mascheroni showed that the use of a ruler was not necessary for these geometrical constructions and that they could be done by the use of compasses alone. His method in the present case is as follows:

Suppose the given points are A and B. First draw the circles with centres A and B and radius BA, with the same radius mark off the points D and E, starting from the intersection C. We know from the usual method of constructing a hexagon that E will be in line with A and B.

With centre E and radius EA, construct an arc to cut the circle with centre A in F and G. Now draw the two arcs with centres F

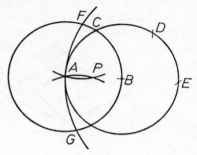

FIG. 7.18

and G and radii FA. These pass through A and a second point, P say. Then P is the required point.

The next four problems have their solutions given, but a knowledge of secondary school mathematics is necessary. They are included for the benefit of the enthusiast, and the solutions include some interesting methods. If you are hardly up to this standard and find difficulty with these, miss them out and resume with the last two problems of this chapter.

The Angle Bisectors Given a triangle ABC which is isosceles with $AB = AC$, prove that the bisectors of angles B and C, drawn to meet the opposite sides in E and F, are equal in length (i.e. $BE = CF$).

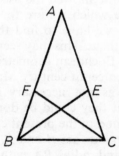

FIG. 7.19

Next prove the converse of this theorem, i.e. in any triangle ABC, given that the bisectors of the two angles, drawn to meet the opposite sides in E and F, are equal in length ($BE = CF$), prove that the triangle is isosceles. (*Solution follows.*)

The first part of the problem is a relatively easy exercise in the use of congruent triangles and the solution of the second part only is given here. It is a fairly well-known example of a type of proof which proceeds on the basis of an initial assumption contrary to what is to be proved until a stage is reached which is contradictory to the assumption.

In the present case, assume that angle C is greater than angle B. Since the angles are bisected and $BE = CF$, this implies that BF is greater than CE (Statement 1).

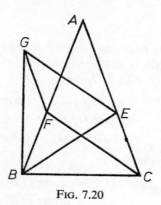

FIG. 7.20

Draw the parallelogram $CEGF$ and join BG.

Since $EG = CF = BE$, triangle EGB is isosceles and angles EGB, EBG are equal. But angle EGF = angle ECF. It follows that

$$\text{angle } EGF > \text{angle } EBF$$
$$\text{angle } FGB < \text{angle } FBG$$
$$FG > FB$$
$$EC > FB$$

This last statement is in contradiction to Statement 1 above. Hence the original assumption is untrue, angles B and C must be equal, and the triangle ABC isosceles.

The Intersecting Chords AB is any chord of a circle, and C is its mid-point. Through C, two more chords DE and FG are drawn. DF cuts AC at H; GE cuts CB at J. Prove $HC = CJ$. (*Solution follows.*)

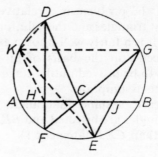

FIG. 7.21

A construction is necessary for the solution of this problem (Fig. 7.21). Through G draw KG parallel to AB to meet the circle at K. Join KD, KC, KH.

Angle KDF = angle KGF		(same segment)
= angle GKC		(isos. triangle)
= angle KCA		(alternate)

Hence $KDCH$ is cyclic, and

$$\text{angle } HKC = \text{angle } HDC$$
$$= \text{angle } FGE$$

It can now be shown that triangles KCH and GCJ are congruent, since

angle HKC = angle FGE	
angle KCH = angle KGC	(proved above)
= angle GCB	(alternate)
$KC = CG$	

The required result follows.

Find the Angle   In Fig. 7.22 ABC is an isosceles triangle with the vertical angle $A = 20°$. Lines BD and CE are drawn as shown, with angle $BCE = 50°$, and angle $CBD = 60°$. Find the size of angle BDE. (*Solution follows.*)

This problem is best solved by trigonometry. It can be easily established that $EB = BC$ and $DB = DA$ from isosceles triangles. Also angle $BDC = 40°$ and $ACB = 80°$. Applying the sine rule to triangle BCD,

$$\frac{BC}{\sin BDC} = \frac{BD}{\sin ACB}$$

$$\frac{BC}{\sin 40°} = \frac{BD}{\sin 80°}$$

$$\frac{EB}{BD} = \frac{\sin 40°}{\sin 80°} = \frac{\sin 40°}{2\sin 40° \cos 40°} = \frac{1}{2\cos 40°}$$

$$= \frac{\frac{1}{2}}{\sin 50°} = \frac{\sin 30°}{\sin 50°} = \frac{\sin 30°}{\sin 120°}$$

By reference to triangle EDB, we now see that angle $EDB = 30°$.

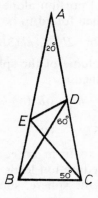

FIG. 7.22

The Hole in the Sphere It is required to bore a hole through a sphere so that the volume of material remaining at the end is just one half of the original volume of the sphere. (*Solution follows.*)

Probably the easiest method of solution uses calculus, and readers familiar with this subject will find a solution in the 'Hints' in Chapter 14 (p. 177). The following solution uses trigonometry.

The formula for the volume of the 'cap' of a sphere (i.e. a section sliced off) is $\frac{1}{6}\pi h(3R^2 + h^2)$, where R is the radius of the cap and h is its height. If Fig. 7.23 represents our problem with AB and CD the sides of the hole, it can be seen that the part cut away is made up of a cylinder $ABCD$ together with a cap above

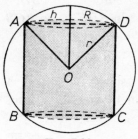

FIG. 7.23

and another cap below it. If r is the radius of the sphere, the height of the cylindrical portion alone will be $2r - 2h$, and the volume of the hole is then found to be

$$\pi R^2(2r - 2h) + \tfrac{1}{3}\pi h(3R^2 + h^2)$$

Since this is half the volume of the sphere it equals $\tfrac{2}{3}\pi r^3$. This equation may be simplified to

$$h(3R^2 + h^2) + 6R^2(r - h) = 2r^3$$

or

$$h^3 + 6R^2r - 3R^2h - 2r^3 = 0$$

(N.B. elimination of h to form an equation in R and r now seems the obvious procedure but leads to difficulties.)

If O is the centre of the sphere, suppose angle AOD to be 2θ. Then

$$\frac{R}{r} = \sin\theta \quad \text{and} \quad \frac{r - h}{r} = \cos\theta$$

or

$$R = r\sin\theta \quad \text{and} \quad h = r(1 - \cos\theta)$$

Substituting in the earlier equation,

$$r^3(1 - \cos\theta)^3 + 6r^3\sin^2\theta - 3r^3\sin^2\theta(1 - \cos\theta) - 2r^3 = 0$$

By the use of $\sin^2\theta = 1 - \cos^2\theta$, this now becomes

$$2r^3 - 4r^3\cos^3\theta = 0$$

Dividing by $2r^3$,

$$\cos^3\theta = \tfrac{1}{2}$$

We may obtain $\sin\theta$ from this by the use of cosine and sine tables, or by using the relation between $\sin^2\theta$ and $\cos^2\theta$ given above. We find $\sin\theta = 0{\cdot}6083$, so that $R = 0{\cdot}61\ r$ (approx.).

Another Hole This problem also concerns a hole to be cut the centre of a sphere, but this time the hole has to be cut in such a way that it is just six inches long.

What is the volume of the sphere remaining after the hole has been cut out? (*Hint p.* 178. *Answer p.* 192.)

We give no worked solution to this problem, which may be solved by calculus or by ordinary mensuration (using the formula for the volume of a 'cap' of a sphere). But those who know nothing of these methods may still obtain the most elegant solution by considering the solution to the problem *Tangenital Chord* given at the beginning of this chapter.

A Final (W)angle! You are given a cube with P, Q and R as three of the vertices (see Fig. 7.24). PQ and PR are the diagonals on two of the faces of the cube. What is the angle between the two diagonals PQ and PR? (*Hint p.* 178. *Answer p.* 192.)

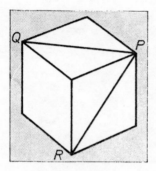

Fig. 7.24

CHAPTER 8

Interlude 2: Cross-number Puzzles

The following cross-number puzzles are constructed in the same manner as ordinary cross-word puzzles, but numbers are substituted for words. The clues may call for calculation or the same type of reasoning which one finds in cross-word puzzles. In all cases, one digit occupies one square in the solution. (*Answers pp.* 192–3.)

Cross-number Puzzle 1

CLUES ACROSS

1 x if $3x+7 = x+41$.
3 $\frac{7}{8}$ of $70\frac{6}{7}$.
5 A perfect cube.
9 Fourth term of series $2, 5, 8 \ldots$
11 $a^2 - b^2$ when $a = 11\frac{1}{2}$, $b = 9\frac{1}{2}$.
12 $7\cdot2 \div 0\cdot3$.
13 $9^{3/2}$
15 x if $21/91 = 3/x$.
16 3 lb in ounces.
17 2^6.
18 Square root of 324.
19 Approx. 7π.
21 $\frac{3}{4}$ of a year in weeks.
23 The largest four-figure square.
25 Positive root of
    $x^2 - 11x - 80 = 0$.
26 $9\frac{3}{4} \div \dfrac{3}{20}$.

CLUES DOWN

2 Five-sixths of a right angle.
3 Square root of 3844.
4 Half of 3 across.
6 The tens digit is twice the units digit.
7 In binary this is 100000.
8 A perfect square.
10 Four terms of a geometric progression with common ratio 2.
12 $111 \times 23 - 5 \times 6 \times 7$.
14 Half a gross plus half a dozen.
15 A quarter of 8 down and still a perfect square.
18 Weeks in a quarter.
19 Highest common factor of 84 and 196.
20 25% of 5lb in ounces.
22 The largest two-digit prime.
23 Lowest common multiple of 32 and 24.
24 Same as 25 across.

Cross-number Puzzle 2

CLUES ACROSS

1 Three times 7 across.
3 Arithmetic progression for the three of us.
5 Put ml for mile.
7 How square can this puzzle be?
9 A teenager in the last of his prime.
10 A fourth power.
11 Divine?
12 He got a fifth more than my score.
14 These four gathered no moss.
16 As far as digits go, there is nothing between 6 and 7.
18 Rearrange 9 down, but still a perfect cube.
19 A quarter of M.

CLUES DOWN

1 My first two digits are the square of my last.
2 An old time record.
3 A long years' war.
4 Emergency?
6 All keyed up on the piano.
8 1 down is only a sixth of it.
9 The only cube in this century.
12 One prime sums the other two primes.
13 I hope my degrees are right.
15 This square mile is good acreage.
16 Reverse my digits and add: result 6 down.
17 $a^b \times b^a$.

Cross-number Puzzle 3

CLUES ACROSS

1   Reversed odd digits.
3   Not quite 3 down.
6   Once a year.
7   The Romans reversed with vim.
9   Just a quarter.
10 The last digit is half the other two.
12 Halve 7 across.
14 Add to your troubles by reversing 1 across.
15 Six times the sum of its digits.
16 Power of two.
17 Consecutive.
19 One short cube.
22 One gross error.
24 Count the yards not the links.
25 All in a foot cube.
28 7^2-2^3.
29 Another right will cause a revolution.
30 Twice a perfect square.

CLUES DOWN

1   Many feet have trodden this mile.
2   Seven in binary notation.
3   Multiple of nine.
4   A cricketer's pair.
5   Add a fifth to this percentage to get full marks.
6   The same either way.
8   Still consecutive.
11 A parallel division.
13 In binary 100000000.
14 A TV line.
17 Change 101 to base 5.
18 Waltztime.
20 A dozen and four dozen.
21 A perfect square in any base greater than 2.
22 Straighten this out with degrees.
23 Divisible by 7, 5 and 11.
26 Round the hundred with 9 across.
27 Yes!

Cross-number Puzzle 4

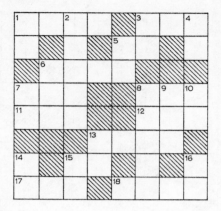

CLUES ACROSS

1 And Harold fell.
3 Why a hundred when it's more.
5 Fathom in feet how deep 'thy father lies'.
6 12345678. That's odd!
7 They are first even with these.
8 Anagramatic 7 across.
11 As each letter is to mile.
12 An unlucky square?
13 That length must be a mile.
15 Approximately seven pies.
17 Here's my bond!
18 This hour's second.

CLUES DOWN

1 Count the letters without the postman.
2 R.F. for O.S.
3 A common base.
4 The timetables have gone by the hour.
5 In his prime between 35 and 40.
6 Not a gross libel.
7 A black dish of song.
8 This knocked eleven for six.
9 Take once a week for five years.
10 Ask darling Clementine.
13 A base duo-decimal figure!
14 Whose score is this?
15 This looks an odd cube!
16 They all make a minute first.

CHAPTER 9

Logicalities

Wotzits A grommit will only fit a sneezle if it has a yatter to interconnect them, but if the sneezle is without its tamble, the grommit can be fitted without the yatter provided that the sneezle is the type that has a nosher, although this type will work just as well with a nosher and a tamble alone as if it is fitted with a grommit. What are the possible working combinations? (*Answer p.* 193.)

'What has this to do with mathematics?' some readers may ask, and on the surface there seems little connection. But the actual calculation is the least important part of a mathematical problem and is often left to machines to perform, e.g. calculating machines, graduated weighing scales, or computers, while the vital part of the problem is its translation into mathematical terms, and the first essential of this process is to sort out the information we have.

Often this information may be available to us in some practical form, as in an engineering problem. At other times it may be contained in some statement—of varying degree of difficulty —which has to be understood first and then operated on. There are various techniques which enable us to deal with problems of the type where the statement seems complex. The simplest method is some form of tabulation.

Staffing Problem In our school, Mr Allen, Mr Baker and Miss Carter teach mathematics, English and French—one to each subject, but not necessarily in that order.

Mr Allen always speaks English when on the Continent.
The French teacher married the English teacher's sister.
Miss Carter borrowed the maths teacher's copy of this book.

Who teaches which subject? (*Solution follows.*)

Construct a table made up of three rows with the three sub-jects, maths, English and French down the side, and three columns headed Mr Allen, Mr Baker and Miss Carter. Now place a 1 or a 0 in any space in the table where according to the statements the teacher at the head of the corresponding column respectively does or does not teach the subject at the start of the corresponding row. Thus the first statement that Mr Allen always speaks English when on the Continent implies that he is not the French teacher and we can insert a 0 in the appropriate space.

	Mr Allen	Mr Baker	Miss Carter
Maths			
English			
French	0		

But the second statement implies that the French teacher is a man, and so must be Mr Baker. We can now insert a 1 in the Baker/French space and also certain 0s in other spaces, since there can only be a single 1 in any row or column.

	Mr Allen	Mr Baker	Miss Carter
Maths		0	
English		0	
French	0	1	0

The third statement implies that Miss Carter is not the maths teacher, so a 0 goes in the Carter/maths space, and the first row may now be completed by putting a 1 in the Allen/maths space. The table may then be completed.

	Mr Allen	Mr. Baker	Miss Carter
Maths	1	0	0
English	0	0	1
French	0	1	0

Hence Mr Allen teaches maths, Miss Carter teaches English and Mr Baker teaches French.

Night Train   It was a foggy night and the engine driver, fireman, guard and sleeping-car attendant of the midnight train met outside my compartment. I could hear their voices but I could not see them. Their conversation went as follows:

'Not only foggy, but cold, so see you get plenty of heat through to my van.'

'Don't blame the driver; blame Edwards if you start to shiver, because he is always playing with the heat regulators.'

'You've done one spell of duty to-night, Brown; what was it like up the line?'

'I didn't think it was too bad, Robinson, but the fellow who was driving didn't like it.'

'Well, come on! It's time we got up front.'

'How do you know which is front, Thomson, in this fog?'

'I just follow the driver as usual!'

What were the respective names of the driver, fireman, guard and attendant? (*Hint p.* 179. *Answer p.* 193.)

Partners Of four couples at the club dance recently, during one dance Joan was dancing with Ann's husband, Alan was dancing with Ann, Bill's wife was dancing with John, Mary's husband was dancing with Alan's wife, and no wife was dancing with her own husband. If Pat was the other wife and Eric the other husband, who was married to whom? (*Hint p.* 179. *Answer p.* 193.)

Professional Problem Bennett, Johnson, Smithers and Woods are (though not necessarily respectively) accountant, solicitor, doctor and dentist. The accountant does Smithers' accounts and gets his medical treatment free as one of the doctor's National Health patients. Johnson does not know Bennett, although his surgery is in the same street as Bennett's office. Johnson and the accountant usually settle each other's bills by mutual adjustment. What are the occupations of each of the four gentlemen? (*Hint p.* 179. *Answer p.* 193.)

The use of a table is not essential for the solution of problems such as these, and you may have been able to solve them by logical deduction from the facts without recourse to the tabulation method. Another way in which logical problems may be solved is by the use of a special type of algebra sometimes called sentence logic.

The puzzles at the end of Chapter 6 used sets to solve certain types of logical problem, and this suggests that we might apply

G

the idea of sets to problems of the type which we are now considering. In fact, the algebra of sentence logic has similar laws and rules to the algebra of sets. This is not quite the same as saying that set algebra and sentence logic are the same thing; what we are saying is that they have very close similarities. In mathematics we say that they are isomorphic, which means that they behave in the same way. There is another isomorphic form in the present case, when we apply a similar algebra to electrical circuits with switches in them. The difference between these three algebras really lies in what the letters represent and what operations are denoted by the different signs.

To illustrate this let us go back to Chapter 6, where we denoted the universal set by U. 'Universal set' means the set which includes all things of the type we are considering, e.g. all the boys in a particular class at school, or all positive integers. The corresponding symbol in sentence logic and in the algebra of circuits is 1. In sentence logic this means the truth of everything which we are saying, or the universal truth; in circuitry it means that an electrical current is flowing. Again, in sets we denote a particular set by a letter; thus A may represent the set of boys in the class who wear spectacles. In sentence logic a letter is used to denote a particular statement, e.g. A means that Alan goes to work on his bicycle. In circuitry letters are used to name switches, A, B, C, etc., in a particular circuit.

In each of our three cases we also have certain operations to perform. In sets, we combine the sets by an operation called union, which is denoted by $\cup$, and an operation called intersection, denoted by $\cap$. In logic the signs $\vee$ and $\wedge$ are used, but to make it easier we shall use the signs we use in circuitry where $+$ and $\times$ are used to denote the operations, and we find a similarity between $+$ and $\cup$ and between $\times$ and $\cap$.

In circuitry $+$ indicates switches in parallel and $\times$ switches in series. In sentence logic $+$ will mean 'or' in the sense of 'either or both' and $\times$ will mean 'and'. Notice how these are the same as the switch requirements in circuitry for the lamp to be lit; in Fig. 9.1 (a) the lamp is lit if either or both switches are on, and in Fig. 9.1 (b) one switch and the other have to be on. If we look at a Venn diagram for two sets A and B which intersect as shown in Fig. 9.2, we see that the intersection is the cross-shaded part and for a member of the sets to lie in this part it must be a member of both A and B, while the union is represented by the

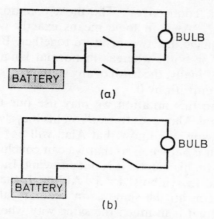

FIG. 9.1 (a) Lamp lights if either or both parallel switches are on; (b) series switches light lamp when both are on.

area shaded in any manner, so that a member of the union of sets A and B lies in set A or in set B or in both.

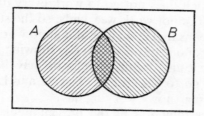

FIG. 9.2

We are now going to investigate sentence logic a little more deeply and then apply it to the solution of some logical problems. The first requirement is to turn given statements into algebra, which we do by using a letter to represent a particular statement. Thus A may denote that Alan goes to work on his bicycle; S may denote the statement that it is snowing. A dash (') is used to express the negation of a particular statement, so that A' means that Alan does not go to work on his bicycle, and S' means that it is not snowing. (Do you remember that we used a similar notation for sets in Chapter 6?)

Just as in ordinary language we connect shorter sentences together to make longer ones, so in logic we link our statements together with 'connectives'. The most common of these are 'and' and 'or'. 'And' in logic means exactly what it does in ordinary language, i.e. two or more together. But 'or' in logic means either or both, whereas in normal language it usually means either. Finally, the truth of any statement(s) is represented by 1 and the untruth by 0.

To illustrate this notation we may use our two statements above: A means Alan goes to work on his bicycle and S denotes that it is snowing. We know that Alan will not go to work on his bicycle if it is snowing, so that we can conclude that if Alan goes to work on his bicycle it is not snowing. In letters we have A and S', and in symbols $A \times S'$. Although the multiplication sign does not mean the same as in ordinary algebra, we find that we can treat it in much the same way when we work out expressions, and also we may omit it from our expressions and use a juxtaposition of letters to mean what is represented by the multiplication sign. For example, in ordinary algebra $3x$ means 'three times x', and in the present case AS' means the same as $A \times S'$.

Obviously Alan does not go to work every day since he has holidays and does not work weekends, so the statement $A' + S$ is true. This means that Alan does not go to work on his bicycle *or* (in the sense of one or the other) it is snowing, as may happen at a weekend; it also means that both may be the case, i.e. Alan does not go to work on his bicycle and it is snowing, as may happen on a snowy working day.

It is not our intention in the present book to cover fully the mathematical theory of all the principles that we shall use; these may be obtained from books mentioned in the bibliography and similar texts. What we wish to do is to obtain an adequate knowledge to solve problems. We shall state that we can apply our normal rules of addition and multiplication to the algebra of sentence logic with certain modifications. Thus in ordinary algebra we know that

$$(a+b)(c+d) = ac+ad+bc+bd,$$

and we may use this in sentence logic so that

$$(A+B)(C+D) = AC+AD+BC+BD.$$

However, in ordinary algebra we also have $a \times a = a^2$, and there are no powers of letters in our new algebra!

Reverting to sets and Venn diagrams, the Venn diagram for set A is shown in Fig. 9.3.

FIG. 9.3

The union of set A with set A (i.e. with itself, denoted by $A \cup A$) is A, and the intersection of set A with set A (i.e. $A \cap A$) is A. Remembering that $\cup$ is similar in operation to $+$ and $\cap$ to $\times$ we have:

$$A + A = A \qquad (1)$$
$$A \times A = A \qquad (2)$$

Also, remembering that the universal set is equivalent to 1, Fig. 9.4 shows that if we have the union of A and A' (A' is the shaded area) it equals the universe, or 1, and the intersection of A and A' is the set with nothing in it, denoted in logic by 0.

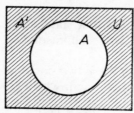

FIG. 9.4

This gives

$$A + A' = 1 \qquad (3)$$
$$AA' = 0 \qquad (4)$$

The above four laws are of great importance in solving logical problems. The following are also of value:

$$1 + A = 1 \qquad (5)$$
$$0 \times A = 0 \qquad (6)$$
$$(A + B)' = A'B' \qquad (7)$$
$$(AB)' = A' + B' \qquad (8)$$

These laws may be confirmed using set algebra and Venn

diagrams. The complete laws and proofs will be found in the appropriate text books.

The following problem appeared in Chapter 1 of this book, and we now solve it using sentence logic.

Suspects Inspector Brown was trying to find the culprit from his three suspects. He knew that it had been a one-man job, and he knew that each suspect would make one true statement and one false statement.

James said: I didn't do it. Thomson didn't do it.
Thomson said: I didn't do it. Frazer didn't do it.
Frazer said: I didn't do it. I don't know who did it.

Who did Inspector Brown arrest? (*Solution follows.*)

Denote the fact that James did it by J, and the fact that he did not do it by J'. Similarly for T and T', and for F and F'. Take the statements in turn and write down the possibilities.

If James' first statement is true and his second false, then Thomson did it and James and Frazer didn't. This can be written $J'TF'$.

Alternatively, if James' first statement is false, his second is true, and James did it, while Thomson and Frazer did not. This is written $JT'F'$.

Since one or other of these two possibilities must be the case, we connect them by the + sign, which gives

$$J'TF' + JT'F'$$

Thomson's statement may be dealt with in a similar manner. This time we obtain

$$J'T'F + J'TF'$$

Frazer's statement is rather different and simply amounts to the fact that he either did it or he didn't! (We are not required to make any logical conclusions ourselves such as the fact that if he did it himself he must know who did it; such matters are taken care of by the algebraic processes.) In Frazer's case we list all the possibilities and connect them, giving

$$JT'F' + J'TF' + J'T'F$$

Finally the statements must be taken together—James'

statement *and* Thomson's *and* Frazer's. Since 'and' in our algebra requires multiplication, we multiply the three expressions, then—since they are simultaneously true—put this equal to 1.

$$(J'TF' + JT'F')\,(J'T'F + J'TF')\,(JT'F' + J'TF' + J'T'F) = 1$$

The equation has now to be simplified and solved using the rules already given. There are various ways of doing this, but probably the easiest is to multiply out the brackets in pairs, making use of the fact that $AA = 1$ and $AA' = 0$. Thus if we multiply the first two brackets, $J'TF'$ times $J'T'F$ becomes 0 since it involves the two terms F and F' (also T and T') which when multiplied become zero. Since anything multiplied by zero is zero, the product as a whole is zero. We next multiply $J'TF'$ by $J'TF'$ which equals $J'TF'$ since A times A is A. Finally we multiply $JT'F'$ by $J'T'F$ and also $JT'F$ by $J'TF'$; in each case obtaining zero. The equation has now reduced to

$$J'TF'\,(JT'F' + J'TF' + J'T'F) = 1$$

which we may multiply out using the same principles as before. This finally becomes

$$J'TF' = 1$$

Hence it is clear that Thomson is the guilty party, and James and Frazer are not guilty.

Copycats   Someone had stolen John's book to copy his homework. He knew it was one of Alan, Bill, Charlie or Denny, and he knew that boys in his class are either wholly truthful or consistent liars. In reply to his questions,

> Alan said, 'Bill hasn't got it. Denny has it.'
> Bill said 'I haven't got it. Charlie has it.'
> Charlie said 'Alan hasn't got it, Bill has it.'
> Denny said 'I haven't got it. I did my own homework.'

John immediately delivered a shrewd and unexpected blow and recovered his book from the recipient. Who had it? (*Hint p.* 179. *Answer p.* 193.)

Some problems are more difficult at the start, when statements have to be turned into algebraic expressions. An example is given below, but it should be noted that once the equation is obtained, the solution follows in the same manner as before.

Evening Out I want to take some friends to the theatre, but each of them is difficult about the way I make up the party. In which different ways can I do this if:

Eric will only go if Alice is there,
Brenda will not go with David unless Eric is there,
Connie won't go with David or Alice,
Alice will only go if Brenda is going,
and I don't want to be the only male in the party?

(Solution follows.)

Take each statement in turn.

(a) Eric will only go if Alice is going.

Taking E to mean that Eric will go, E' to mean he will not, and similarly for the others, we get from the first statement that the presence of Eric implies the presence of Alice, or E implies A. The expression for this is $E' + A$, since this gives the possibilities E' (neither Eric nor Alice there), E' and A (Alice going but not Eric), and A and not E' (notice the double negative here since we are not choosing the possibility of E'; this means Eric may go and Alice also).

(b) Brenda will not go with David unless Eric is there.

This can be phrased that the presence of Brenda and David implies the presence of Eric, or BD implies E. This is the same as (a) above with BD in place of E, and E in place of A, making the expression $(BD)' + E$.

(c) Connie won't go with David or Alice.

The possibilities for David and Alice are represented by $D + A$, since either or both may go, but Connie won't go with any of these combinations, so C implies $(D + A)'$. The final expression is $C' + (D + A)'$. (Check the possibilities as we did in part (a).)

(d) Alice will only go if Brenda is going.

Presence of A implies presence of B, which is similar to condition (a) and the expression in this case is $A' + B$.

(e) I don't want to be the only male in the party.

This means that either David or Eric or both must go, and the corresponding expression is $D + E$.

These conditions may also be dealt with by Venn diagrams; see Fig. 9.5.

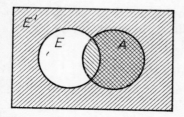

All shaded: $E' \cup A$ or $E'+A$

(a) ERIC WILL NOT GO WITHOUT ALICE

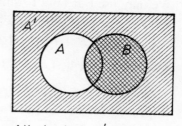

All shaded: $A'+B$

(d) ALICE WILL NOT GO WITHOUT BRENDA

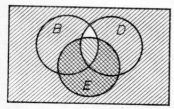

$(B \cap D)'$ or $(BD)'$

E

All shaded: $(BD)'+E$

(b) BRENDA WILL NOT GO WITH DAVID IF ERIC IS NOT THERE

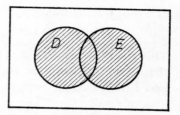

Shaded area: $D+E$

(e) DAVID AND/OR ERIC MUST GO

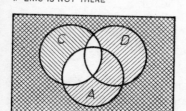

C'

$(D+A)'$

All shaded: $C'+(D+A)'$

(c) CONNIE WON'T GO WITH DAVID OR ALICE

FIG. 9.5

The statements may be illustrated by Venn diagrams: in each case, the total area shaded represents the required condition. Remember that $+$ and $\cup$ correspond and also $\times$ and $\cap$.

Conditions (a) to (e) must be simultaneously satisfied, so we connect them by 'and', i.e. we multiply them together, and we put the whole expression equal to 1. This gives

$$(E' + A) ((BD)' + E) (C' + (D + A)') (A' + B) (D + E) = 1$$

We may use our laws to simplify and solve the equation; thus, using law (7) and law (8) given earlier,

$$(E' + A) (B' + D' + E) (C' + D'A') (A' + B) (D + E) = 1$$

Multiplying the first and last brackets, and noting that $EE' = 0$,

$$(E'D + AD + AE) (B' + D' + E) (C' + D'A') (A' + B) = 1$$

Repeating for the first and last brackets,

$$(A'DE' + BDE' + ABD + ABE) (B' + D' + E) (C' + D'A') = 1$$

Now simplifying the first two brackets,

$$(A'B'DE' + ABD'E + ABDE + ABE) (B' + D' + E) = 1$$

and multiplying out,

$$A'B'DE' + ABD'E + ABDE + ABE = 1$$

This gives the solution that the possible combinations of friends that I can take with me are David by himself, or Alice, Brenda and Eric, or Alice, Brenda, David and Eric.

Committee Quorum Jones, Smith, Macmillan and Riley are eligible to serve on a certain committee, but they stipulate the following conditions:

Jones will only serve if Macmillan serves.
Smith will not serve if Macmillan and Riley serve together.
Macmillan will not serve with Smith unless Jones also serves.
Riley will only serve on the committee if Jones is on it also.

How should the committee be constituted? (*Hint p.* 180. *Answer p.* 193.)

Club Run Ken, Len, Mary and Nancy are all members of the cycling club, but Ken will not go on a run if Len is going. As both are officials, one or other must attend. Mary says she will only go if Ken is going, and both Ken and Len refuse to go alone. On the other hand, Nancy will only go if Mary is going. Who should turn up for the run? (*Hint p.* 180. *Answer p.* 193.)

Illuminations Five coloured electric bulbs are arranged in a special circuit, which switches combinations of them on or off according to a pre-set programme. There are five such programmes:

1. If the white lamp is lit, both the blue and the amber are also lit.
2. If the amber is lit, the red is lit.
3. Either the blue and white are both lit, or one of them is.
4. Blue and Green are both on or both off together.
5. Red and Green may not be lit together.

If all the programmes are switched on at once, which bulbs will be lit? (*Hint p*. 180. *Answer p*. 193.)

Coach Trip Our local coach firm runs evening trips through the prettier villages of the neighbourhood. But because some of the roads are narrow and the time is limited, there are restrictions on the choice of route. The driver knows that if he goes through Appleton he must go through Coppleton. On the other hand, he cannot go through Brabham if he intends to go through both Appleton and Egham. To get to Brabham or Coppleton he has to go through Dickle, and if he goes to Dickle he must go through Brabham or Coppleton, but he cannot go through both. He particularly wants to go through Dickle and Egham since they are the prettiest places. What possible routes can he go? (*Hint p*. 180. *Answer p*. 193.)

We end the chapter with some general logical problems. These involve various methods of solution and are left to the reader's ingenuity. The main feature of all such problems, of course, is the need to sort out the information into a form which is comprehensible and manageable. A process of elimination—by trial and error, by tabulation, or by some other method—then yields a solution. It can be interesting and worthwhile mathematically to try and solve the same problem by different methods, to compare these, and to decide which seems to be the best method of solution.

The Fibs and the Trues There have been various teenager cults, and the latest craze in our district is the division into the Fibs and the Trues. Unfortunately, as one of the older generation, I can never tell the two apart; the only thing is to ask them, but even here there is a difficulty. The Fibs always tell the truth, and the Trues always tell lies. Now the other day I met Paul along with a couple of his friends, John and Pat. I asked Paul whether he was a Fib or a True, but I didn't quite catch what he said because he was laughing at the time. 'What did he say?' I asked the other two. John replied 'He said he was a Fib!' Pat replied 'He said he was a True!' What were John and Pat? (*Hint p.* 180. *Answer p.* 193.)

Abstract Art If *A*, *B*, *C*, and *D* are symbols without any assigned meaning, and *W*, *X*, *Y* and *Z* are similar symbols, establish a relationship between them, given that

If *A* is *X*, then *C* is not *Y*
If *B* is not *Y*, then *D* is *Z*.
If *C* is either *X* or *Y*, *D* is not *W*.
If *D* is *W*, then *C* is *Z*.

(*Hint. p.* 180. *Answer p.* 193.)

Hats Off! Three persons *A*, *B* and *C*, sit in a row with *A* in front, *B* behind him, and *C* behind *B*. Five hats—three white and two red—are placed in a bag, and *A*, *B* and *C* know this. A hat is taken from the bag and placed on *A*'s head, so that *A* does not see what colour it is; *B* and *C* can see, of course, since *A* is in front of them. Another hat is taken from the bag and placed on *B*'s head so that *B* cannot see what colour it is; *C* can see *B*'s hat since he is sitting behind, but *A* cannot see it since he is sitting in front of *B*. Finally a hat is placed on *C*'s head without allowing *C* to see it; *A* and *B* cannot see it, of course. *C* is now asked if he can state the colour of the hat which he is wearing; he is not required to state the colour, but simply to say 'yes' or 'no'. He replies that he cannot. *B*, having heard this reply, is now asked if he can or can not say what is the colour of his hat. He replies that he cannot. From the two replies which *A* has heard, is *A* able to say what is the colour of the hat which he is wearing?

This problem may now be extended. Consider the various combinations of hats which may be worn; how many combinations are there? Is it possible for *A* to answer correctly after a similar procedure in each case?

Finally, consider the general case of *n* persons with *n* hats of one colour and *n* + 1 hats of another colour. (*Answer p.* 193.)

CHAPTER 10

And So On . . . And So Forth . . .

Find the sum of the whole numbers from one to twenty inclusive.

What do you do? Add one and two, then add three to the total, then four and so on? Or do you arrange them like this:

$$
\begin{array}{ll}
1+20 & 21 \\
2+19 & 21 \\
3+18 & 21 \\
4+17 & 21 \\
5+16 & 21 \\
\text{and so on.} &
\end{array}
$$

How many rows? So the total is . . .?

In Chapter 3 we looked at patterns in mathematics and examples like the one above establish patterns with numbers so that we are able to continue the pattern and also find the sum of a certain number of terms in this pattern. A succession of related numbers like this is called a series.

Probably the two most common series are arithmetic series and geometric series, often referred to as arithmetic progressions (A.P.) and geometric progressions (G.P.). Most algebra text-books have sections on these, but as we so often need to sum these series we will consider their summation here.

The numbers from one to twenty form an arithmetic progression with a common difference of one, i.e. the difference between successive terms is one. Suppose we have such a series with a difference of d, and the first term of the series is a. Then each term after the first consists of a together with a certain number of d: the second term is $a+1d$, the third term is $a+2d$, the fourth $a+3d$, and so on, to the nth term which is $a+(n-1)d$. Thus the sum of n terms of the series may be written

$$
S = a+(a+d)+(a+2d)+(a+3d)+ \ldots +(a+(n-1)d)
$$

Noting what we did with the numbers 1 to 20, we re-write this sum as

$$S = \left(a+(n-1)d\right)+\left(a+(n-2)d\right)+ \ldots +(a+d)+a$$

If we add the corresponding terms of each series, the first, the second, the third, etc., we obtain $2a+(n-1)d$ each time and there are n such terms. Hence

$$2S = n\left(2a+(n-1)d\right)$$

and so

$$S = \tfrac{1}{2}n\left(2a+(n-1)d\right)$$

Arithmetic Progression Use the formula to obtain the sum of all the even numbers from 1 to 50, and all the multiples of three between 100 and 200. (*Hint p.* 180. *Answer p.* 193.)

A geometric progression is a series where each successive term is *multiplied* by a particular number (in this case called the common ratio). Thus 2, 6, 18, 54 . . . is a geometric series with common ratio three, since any term is three times the preceding one. For this series, if a represents the first term and d the common ratio, the sum of n terms of the series is

$$S = a+ar+ar^{2}+ar^{3}+ar^{4}+ \ldots +ar^{n-1}$$

Now multiply each side of this equation by r and write the two series below each other, like this:

$$S = a+ar+ar^{2}+ar^{3}+ \ldots +ar^{n-1}$$
$$rS = \quad\; ar+ar^{2}+ar^{3}+ \ldots +ar^{n-1}+ar^{n}$$

Subtract the top from the bottom series

$$rS-S = ar^{n}-a$$

Whence

$$S = \frac{a(r^{n}-1)}{r-1}$$

(This formula was used in a problem in Chapter 1 of this book.)

Clearly, both arithmetic and geometric series may continue indefinitely by making n infinitely large. If you take an infinite number of terms of an arithmetic progression such as $2+4+6+8+ \ldots$, the sum will be infinite, but in the case of the geometric series, provided r is less than one, the sum of the series approaches a particular value, or limit. As an example, we may quote the series

$$\frac{1}{2}+\frac{1}{4}+\frac{1}{8}+\frac{1}{16}+\frac{1}{32}+ \dots \text{ to infinity}$$

This sum may be represented by taking a square, the area of which is regarded as unity, then dividing it into two equal parts, dividing one of these parts into two equal parts, halving one of these, and so on. Each successive part represents a term of the series; it can be seen that the process can be continued indefinitely; and that the sum of the whole series is one.

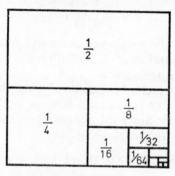

FIG. 10.1

Alternatively, of course, the formula proved above may be used to sum the series. In the present case $a = \frac{1}{2}$ and $r = \frac{1}{2}$. When r is fractional and n is infinite, r^n becomes zero, so the formula becomes

$$S = \frac{a}{1-r}$$

In this particular case,

$$S = \frac{\frac{1}{2}}{1-\frac{1}{2}} = 1$$

We can only take this sum to infinity when r is less than one. If r is greater than one, as in the series $3+6+12+24+ \dots$, each successive term becomes larger and the sum of an infinite number of terms becomes infinity.

On the Bounce   A ball is dropped from a height of ten feet. Each time it bounces it rises to two-thirds the height from which it fell before that bounce. How far does the ball travel before it comes to rest? (*Solution follows.*)

The motion may be represented in diagram form, as shown in Fig. 10.2.

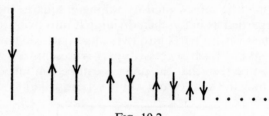

FIG. 10.2

In theory at least the motion goes on indefinitely, the bounces becoming smaller and smaller, but still continuing. So we obtain a series and sum it to infinity.

The height to which the ball rises after the first bounce is $10 \times \frac{2}{3}$ feet, that after the second bounce is $(10 \times \frac{2}{3}) \times \frac{2}{3}$ feet or $10 \times (\frac{2}{3})^2$ and so on. Since all the bounces, except the first fall, are up and down, the total distance is

$$10 + 2 \times 10 \times \tfrac{2}{3} + 2 \times 10 \times (\tfrac{2}{3})^2 + \dots \text{ to infinity}$$
$$= 10 + 20 \times \tfrac{2}{3} + 20 \times (\tfrac{2}{3})^2 + \dots$$
$$= 10 + 20 \, [\tfrac{2}{3} + (\tfrac{2}{3})^2 + \dots]$$

The series in the square brackets is an infinite series with first term $\frac{2}{3}$ and common ratio $\frac{2}{3}$, which we sum. Hence the total distance is

$$10 + 20 \times \frac{\frac{2}{3}}{1 - \frac{2}{3}} = 10 + 40 = 50 \text{ feet}$$

Many problems on series originate from repeating a process or action, as in the following examples.

Line Intersections Draw two straight lines to intersect, then another straight line to intersect these (not at the same point as before). Make a table of the number of lines and the number of intersections.

No. of lines	No. of intersections
2	1
3	3

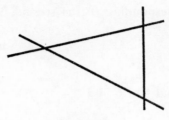

FIG. 10.3

Draw another line to cut these in different points, then another; in each case list the number of lines and intersections. Without drawing a further line, see if you can fill in the next row of the table. Then try the succeeding row. Once you have achieved this, see if you can fill in the number of intersections when the number of lines is *n*. (*Solution follows.*)

The table should continue

| 4 | 6 |
| 5 | 10 |

If you found this pattern without drawing the lines you have established the relationship between the number of lines and intersections, and it is interesting to see in how many different ways this relationship may be developed. A fairly obvious one is by adding the number of lines and number of intersections in any row, and this gives the number of intersections in the following row. Thus, in the first row $2+1 = 3$, which is the number of intersections in the second row, and in the second row $3+3 = 6$, the number of intersections in the third row, etc. Unfortunately this method does not allow any inference of the number of intersections for *n* lines.

Another method is to consider the problem geometrically, and to say that since each successive line must cut each of the previous lines, then each time we must add the number of lines already drawn to the existing total. Thus when we add another line to the existing three lines, we shall have three more intersections, making a total of six. This, however, is simply a geometric way of describing our first method.

A less obvious, but more fruitful way, is to notice that if we multiply each successive number of lines, this is just twice the

H

intersections corresponding to the second of these two rows. Thus—

Lines	Intersections
2	
3	3
$2 \times 3 = 6$ and $\frac{1}{2}$ of $6 = 3$ _____↑	
3	
4	6
$3 \times 4 = 12$ and $\frac{1}{2}$ of $12 = 6$ _____↑	

Or, in other words, the number of intersections on any row is half the product of the number of lines on that row and the number of lines on the previous row. By continuing this pattern for the rows containing $n-1$ and n lines we have

$n-1$
n $\frac{1}{2}n(n-1)$
$n \times (n-1) = n(n-1)$ and $\frac{1}{2}$ of $n(n-1) = $ ____↑

Therefore the number of intersections for n lines is $\frac{1}{2}n(n-1)$.

Circle Intersections Proceeding as in the previous example, find the number of intersections when 2, 3, 4, ..., n circles intersect. (*Answer p.* 193.)

Space Problem Repeat the line intersection problem, but this time, instead of listing and forecasting the number of intersections of the lines, count and forecast the number of spaces contained between the lines. (*Hint p.* 181. *Answer p.* 194.)

Triangular Numbers Triangular numbers are arrangements of dots or crosses to form a triangle. The first few numbers are shown below.

```
    +           +            +                +
       + +         + +            + +
                + + +          + + +
                            + + + +
    1           3            6                10
```

Find how this pattern is built up and find a formula for the nth triangular number. Then find the sum of n triangular numbers. (*Solution follows.*)

The clue to this problem lies in the pattern of crosses. If, for each term, we take the existing pattern and add a similar pattern (represented below by the use of 'o's) we can make each number pattern into a rectangle; thus for six and ten:

```
o  o  o      o  o  o  o
+  o  o      +  o  o  o
+  +  o      +  +  o  o
+  +  +      +  +  +  o
             +  +  +  +
```

The number of symbols in each of these rectangles conforms to a pattern: the first contains 4×3 symbols and is derived from the third triangular number, while the second contains 5×4 symbols and is from the fourth triangular number. Reverting to the crosses we see that the third triangular number is $\frac{1}{2} \times 3 \times 4$, the fourth is $\frac{1}{2} \times 4 \times 5$, and proceeding similarly the nth triangular number will be $\frac{1}{2}n(n+1)$. The sum of n triangular numbers is left for you to find.

Pyramid Numbers   Take the simple arithmetic series with first term 1 and common difference 1, thus:

$$1, 2, 3, 4, 5, \ldots$$

Add the terms from the first to each term successively, to obtain the series

$$1, 3, 6, 10, 15, \ldots$$

(Where have we met this before?) Now take the series with first term 1 and common difference 2:

$$1, 3, 5, 7, 9, \ldots$$

Add terms as before. What series of numbers do we obtain? Can you represent them by series of dots or crosses on a paper? (*Answer p.* 194.)

(The process could be continued, of course, using the series 1, 4, 7, 10 . . . or 1, 5, 9, 13, . . . etc. as starting point.)

Series such as the one obtained above, and also triangular numbers, are called figurate numbers, the arithmetic series being known as linear figurate numbers. It is now possible to develop into three-dimensional representations.

Starting with the triangular numbers

$$1, 3, 6, 10, 15, \ldots$$

sum successively in the same manner as before to obtain the series

$$1, 4, 10, 20, \ldots$$

These are pyramid numbers. Structure these in terms of a solid figure of billiard balls, piles of tins in a supermarket, or some similar arrangement. The first number represents one tin (or billiard ball) by itself, the second number the pile formed by putting this one tin on top of three others, the third number represents the tins in a pile with three layers containing one, three and six tins in successive layers, and so on (see Fig. 10.4). Find the nth term of the above series. (*Hint p.* 181. *Answer p.* 194.)

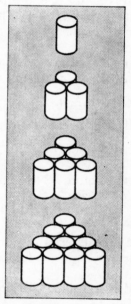

FIG. 10.4

It is interesting to proceed, by the same process, from the third dimension to the fourth, and from that to the fifth and higher dimensions of figurate numbers. Since space is three-dimensional, it is difficult to visualize n-dimensional structures to correspond with the above results; this is an example of the way in which mathematics may extend beyond our physical world.

In the above examples, we added terms to form a series, and this is reminiscent of Pascal's triangle and of the Fibonacci series, both well-known mathematical patterns. Pascal's triangle starts with the figure 1, and obtains successive rows of the pattern by adding pairs of terms from the row above. Each row may be assumed to have 0 at each end, so that the first row would be regarded as 0 1 0, the second row as 0 1 1 0, the two 1s being obtained by adding $0+1$ and $1+0$ in the row above. The third row comes from the addition of $0+1$, $1+1$, $1+0$, and similarly for the following rows. The first few rows are as follows:

										n
				1						
			1		1					1
		1		2		1				2
	1		3		3		1			3
1		4		6		4		1		4
1	5	10	10	5	1					5
1	6	15	20	15	6	1				6
1	7	21	35	35	21	7	1			7
1	8	28	56	70	56	28	8	1		8

The numbers in each row of the Pascal triangle represent the co-efficients of the expansion of $(1+x)^n$ when n has, in each case, the value given alongside in the above table. Thus

$$(1+x)^2 = 1+2x+x^2$$
$$(1+x)^3 = 1+3x+3x^2+x^3$$
$$(1+x)^4 = 1+4x+6x^2+4x^3+x^4$$

and so on.

The Fibonacci series is obtained by writing the figure 1 twice, and then generating successive terms by adding the two previous terms of the series. This gives

1, 1, 2, 3, 5, 8, 13, 21, 34, 55, . . .

Each of these results—Pascal's triangle and the Fibonacci series—have many interesting applications, and it is not possible to list all these here, but the following examples indicate a few of their properties.

Figurate Numbers Remembering the earlier work of this chapter, what connection have figurate numbers with Pascal's triangle. Can you find a reason for this? (*Answer p.* 194.)

Street Plan In a certain city where the streets are arranged on a rectangular plan (i.e. streets in one direction intersect at right angles the streets running in the other direction), a man start to walk from a certain corner A to another corner B six streets away from A in one direction and five streets from A in the other direction. (Backward moves are not allowed.) How many different routes may the man take to reach his destination? Generalize the result to give the number of ways in which he may reach the intersection of the pth street in one direction with the qth street in the other direction (*Hint p.* 181. *Answer p.* 194.)

Across the Chessboard How many different ways are there of moving from one corner of a chessboard to the opposite corner, if moves are made one square at a time in a forward direction

    (a) moving horizontally and vertically only,
    (b) when horizontal, vertical and diagonal moves are allowed?
                            (*Hint p.* 181. *Answer p.* 194.)

Fibonacci Sum Find the sum of the first n terms of the Fibonacci series. (*Solution follows.*)

The solution of this problem employs a method which is often used in summation of series. Each term is expressed as the difference of two terms, and when we sum the series most of the differences nullify each other.

The Fibonacci series is 1, 1, 2, 3, 5, 8, 13 . . . and the rth term may be obtained by taking the $(r+1)$th term from the $(r+2)$th term. We have successively

$$1 = 2 - 1$$
$$1 = 3 - 2$$
$$2 = 5 - 3$$
$$3 = 8 - 5$$
$$5 = 13 - 8$$

and so on until

$$f_{n-1} = f_{n+1} - f_n$$
$$f_n = f_{n+2} - f_{n-1}$$

(where f_n denotes the nth term of the series)

Now add the respective sides of the equations:

$$1 + 1 + 2 + 3 + 5 + 8 + \ldots + f_{n-1} + f_n = f_{n+2} - 1.$$

All the other terms of the right-hand side cancel. The result may be stated: the sum of the first n terms of the Fibonacci series is one less than the $(n+2)$th term.

If we arrange the terms of the Fibonacci series as fractions in turn, and reduce them to decimals, we have the following result:

$$\frac{1}{1} = 1 \cdot 0 \qquad\qquad \frac{2}{1} = 2 \cdot 0$$

$$\frac{3}{2} = 1 \cdot 5 \qquad\qquad \frac{5}{3} = 1 \cdot 666 \ldots$$

$$\frac{8}{5} = 1 \cdot 6 \qquad\qquad \frac{13}{8} = 1 \cdot 625$$

$$\frac{21}{13} = 1 \cdot 6154 \ldots \qquad\qquad \frac{34}{21} = 1 \cdot 61905 \ldots$$

The results in the first column are gradually increasing while those in the second are decreasing, and if the process is continued it will be found that each of the two columns approaches the figure $1 \cdot 618034 \ldots$ This figure is the value of $(1 + \sqrt{5})/2$.

The more adventurous reader might now try his hand at finding the formula for the nth term of the Fibonacci series. The exercise is not an easy one, nor is the result, since the fraction just given features in the solution.

Skew Pascal Triangle   The following property is obtained from the Pascal Triangle, but is more easily seen if we re-write it as shown:

```
                                              1
                                        1
                                  1           1
                            1           2
                      1           3           1
                1           4           3
          1           5           6           1
    1           6          10           4
1         7          15          10           1
    8          21          20           5
    9          28          35          15           1
```

and so on.

Now add each row of figures horizontally. What do you notice? Can you give the reason for this? (*Answer p.* 194.)

Cube It Take any number of consecutive digits starting from 1 and add them. Then find the cubes of each of these same digits and sum them. Compare the results and explain. (*Answer p.* 194.)

Split the Circle A circle is divided into two semi-circles, each being half the diameter of the large circle, so that the large circle is divided by them into two equal parts.

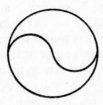

FIG. 10.5

Suppose, however, that after drawing the first semi-circle, we draw the next one only half the diameter of the first, and then continue to draw semi-circles on alternate sides of the diameter, each time making the semi-circles half the diameter of the previous one (Fig. 10.6).

What is the ratio now of the two parts into which the main circle is divided? (*Hint p.* 182. *Answer p.* 194.)

FIG. 10.6

Varied Descent A building under construction has three rows of vertical steel girders with three girders in a row, and these are interconnected by horizontal girders on every one of three floors, as shown in Fig. 10.7.

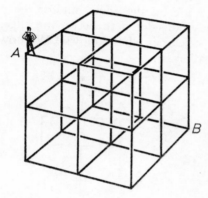

FIG. 10.7

A man may descend by walking along any horizontal girder and climbing down any vertical one. If he is standing at one of the top corners (*A*) of the building, how many different routes may he follow in order to reach the diagonally opposite corner (*B*) at the bottom of the building? Backward moves are not allowed. (*Hint p. 182. Answer p. 194.*)

An Economical Ruler This problem differs from the foregoing examples on series. A twelve-inch ruler is to be marked so that it may be used to measure all the integral lengths from 1 to 12 inches. What is the minimum number of markings possible, and which are they? (*Answer p. 194.*)

Overbalance Is it possible to take a number of bricks, all the same size, and pile them up one above the other until the top brick lies wholly outside the area on which the bottom brick rests? (*Solution follows.*)

A point to note in this solution is that the mathematical analysis must be in the opposite direction to the practical method of brick building, i.e. mathematically we must start with the top brick. Starting with the bottom brick, putting another on top of it, and so on, can lead to difficulties.

Consider the top brick placed on the brick below so that it is just about to fall off. The middle of the top brick must be exactly above the edge of the brick below. (For balance, of course, the half-way mark must be just a shade over the lower brick and not quite on the edge, but for the purpose of our problem we may consider it to be on the edge, and similarly with the other bricks in what follows.) We now have the two bricks as shown below:

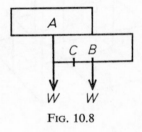

Fig. 10.8

We may imagine the weight of the top brick acting vertically downwards through its centre A, the weight of the lower brick acting downwards from *its* centre B, so that the weight of the two bricks together acts mid-way between the verticals through B and A, at C. BC is one-quarter of the length of the brick, so C must be immediately over the edge of the next lower brick.

Now consider the three bricks together. The weight acting down through C is that of two bricks, and the weight of the third brick acts at its centre point, so that the centre of gravity of the three bricks must act through E, where E divides the distance between the two vertical forces in the ratio $1:2$. In other words, FE is one-sixth of the length of a brick.

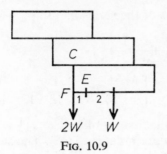

FIG. 10.9

We now see a pattern emerging:

The first overhang is $\frac{1}{2}$
The second overhang is $\frac{1}{4}$
The third overhang is $\frac{1}{6}$
The fourth overhang is $\frac{1}{8}$

and so on. The total overlap for a pile of bricks arranged in this manner will be according to the following series:

$$\frac{1}{2}+\frac{1}{4}+\frac{1}{6}+\frac{1}{8}+ \dots$$

or

$$\frac{1}{2}(1 +\frac{1}{2}+\frac{1}{3}+\frac{1}{4}+ \dots)$$

The series within the brackets can be seen to total more than two for the first four terms, or in other words, the first four terms of the initial series has a total which is greater than one. This means that the fourth brick can completely overhang the bottom one.

The series $1 +\frac{1}{2}+\frac{1}{3}+\frac{1}{4}+ \dots$ is a divergent series, i.e. it never approaches a limit. So its sum to infinity is infinite. What does this mean in terms of building bricks as suggested in the question?

The divergency of the series in the last paragraph may not be apparent immediately. After all, we have already found the series $\frac{1}{2}+\frac{1}{4}+\frac{1}{8}+\frac{1}{16}+ \dots$ to be convergent, and it would seem a reasonable assumption that any series whose successive terms decreased was a convergent series, but this is not so. The divergence of the series $1 +\frac{1}{2}+\frac{1}{3}+\frac{1}{4}+ \dots$ may be shown as follows.

Group the terms of the series like this:

$$1 + \frac{1}{2} + \left[\frac{1}{3} + \frac{1}{4}\right] + \left[\frac{1}{5} + \frac{1}{6} + \frac{1}{7} + \frac{1}{8}\right] +$$

$$+ \left[\frac{1}{9} + \frac{1}{10} + \frac{1}{11} + \frac{1}{12} + \frac{1}{13} + \frac{1}{14} + \frac{1}{15} + \frac{1}{16}\right] + \cdots$$

Now the terms in the first set of brackets $(\frac{1}{3} + \frac{1}{4})$ are greater than $(\frac{1}{4} + \frac{1}{4})$; hence they are greater than $\frac{1}{2}$. The sum of the terms in the second bracket is greater than $(\frac{1}{8} + \frac{1}{8} + \frac{1}{8} + \frac{1}{8})$, and so is also greater than $\frac{1}{2}$. Proceeding similarly with successive sets of terms in each of the brackets we may show that the initial series is greater than $1 + \frac{1}{2} + \frac{1}{2} + \frac{1}{2} + \frac{1}{2} + \cdots$ But this series obviously diverges to infinity, since the more terms we take, the greater is the sum of the terms. So the original series $1 + \frac{1}{2} + \frac{1}{3} + \frac{1}{4} + \frac{1}{5} + \cdots$ is also divergent; it is known as the harmonic series.

On the other hand, the series

$$1 - \tfrac{1}{2} + \tfrac{1}{3} - \tfrac{1}{4} + \tfrac{1}{5} - \tfrac{1}{6} + \cdots$$

is a convergent series. Its sum to infinity is a non-terminating decimal, $0{\cdot}693147\ldots$, and if we denote this by S, the Sth root of two is another important series which is usually denoted by e. In other words, $e^S = 2$. The exponential series e is

$$1 + \frac{1}{1!} + \frac{1}{2!} + \frac{1}{3!} + \frac{1}{4!} + \cdots$$

where 4! means $4 \times 3 \times 2 \times 1$, and similarly for the other terms. The exponential series is also a convergent series whose sum is the non-terminating decimal $2{\cdot}71828\ldots$ It occurs in higher mathematics, e.g. as the theoretical base of logs. The interested reader can find more about this series in advanced algebra books (see Bibliography).

The Expedition Five explorers set up their base camp on the edge of a desert and intend to explore across the desert and return to their base. Each man can carry enough food to last one man for six days. They arrange that they shall set off together, but return singly on different days, each man having enough food to see him back to the base camp; in this way the last man to return will travel as far as possible into the desert.

How many days after the start of the expedition does the last man return to base?

The problem may then be generalized as follows: If there are n explorers, and each man is able to carry enough food to last one man for d days, how many days is it after the start of the expedition when the last man returns to base? (*Hint p.* 183. *Answer p.* 194.)

Second Expedition There are n explorers who start, as before, from their base on the edge of the desert, and each man can carry enough food to last one man for d days. On this occasion, however, they are able to deposit food in subsidiary camps along the route. Again, they arrange to return singly on different days, each taking with him enough food to see him safely back to the base camp, so that again one man can penetrate for the furthest possible distance into the desert. How many days after the start of the expedition does this man return to base? (*Solution follows.*)

The best method is to consider first the last man at the camp furthest from the start. Since the total amount of food that he can carry will last him d days, he can travel outwards from the camp for a time of $\frac{1}{2}d$ days, and return over the same period. Initially we will consider only the outward journey, since the total time will be twice this. When the last man arrives back at the last camp there must be sufficient food there to last him for the time to get to the penultimate camp, i.e. the camp next to the furthest one (see Fig. 10.10). But as he required

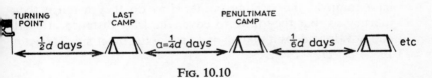

FIG. 10.10

the total amount of food he could carry for his final journey, the stock of food at the last camp, together with the food he consumed going outwards from the penultimate to the last camp, must have been carried by his companion on this leg of the journey. In addition, of course, this companion had to carry food for himself from the penultimate to the last camp and back again. If the time between these camps is denoted by a then the food which the companion carries must be sufficient to feed two men for a days on the outward journey and for a days on their return. Hence $4a$ is equivalent to d; or in other words, the actual time between the last two camps is $\frac{1}{4}d$.

Dealing with the next previous camp in the same way, we see that the food carried by one man has to provide for three men outward and three men inward, so that the time between this camp and the penultimate one is $\frac{1}{6}d$. And so on for the other camps.

Summing these times in one direction only, the time taken by the man who travels furthest to go from the starting point to his turning point is

$$\frac{1}{2}d+\frac{1}{4}d+\frac{1}{6}d+\frac{1}{8}d+ \ldots +\frac{1}{n}d$$

Hence the total time taken for this man to go out to his furthest point and return to the base camp is twice the above series, or

$$d\left(1+\frac{1}{2}+\frac{1}{3}+\frac{1}{4}+ \ldots +\frac{2}{n}\right)$$

Third Expedition Three men set off to cross a desert from one side to the other; all of them are to complete the journey, the distance across the desert being 180 miles. They can travel 20 miles a day, and each man can carry a six-day food supply for one man. They may establish food dumps at various camps across the desert, and there is a plentiful supply of food at their base camp at the starting point. How do they arrange their journey so that they have to cover the least distance in total, and how long does it take them to reach the other side? (*Hint p.* 183. *Answer p.* 194.)

Strictly Numerical 1 Five numbers are in geometrical progression. Show that their product is equal to the fifth power of the middle number.

Generalize the result to show that the product of any odd number of terms of a GP is equal to the rth power of the middle term, where r is the number of terms. (*Hint and Answer p.* 184.)

Strictly Numerical 2   If the common ratio of a GP is less than a half, show that each term will be greater than the sum of all the terms that follow it. (*Hint and Answer p.* 184.)

Strictly Numerical 3   Three successive terms of an arithmetic progression (AP) have the same first term as three successive terms of a GP, and the common ratio of the GP is the same as the common difference of the AP. Also the last term of the GP is the same number of times the last term of the AP as the common ratio (or common difference). If all the terms and the difference and ratio are all integers, show that there are only two possible solutions of this problem, and find the two solutions. (*Hint p.* 184. *Answer p.* 194.)

Geometric Growth   A water lily grows on the surface of a pond so that the area which it covers each day is double the area that it was the previous day. If it takes ten days to cover the entire surface of the pond how long does it take to cover half the surface? (*Answer p.* 194.)

CHAPTER 11

Topological Topics

In Chapter 6 we solved problems by the use of Venn diagrams, in which a closed curve—often a circle—represented a particular set. Since we had intersecting sets, we had intersecting curves. Thus for three sets we used a diagram such as Fig. 11.1.

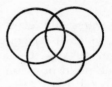

FIG. 11.1

One problem in Chapter 6 involved four intersecting sets (*Language Laboratory*), and it was suggested that Fig. 11.2 will meet that case.

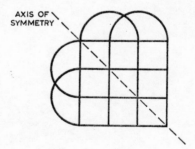

FIG. 11.2

How does one check whether this is so? The spaces in the diagram represent set intersections, so we have to list possible combinations of sets. In the case of three sets we must have sets by themselves, combinations of two sets and combinations of three sets; in total this comes to seven possibilities, so that the Venn diagram must have seven spaces.

The calculation is simplified when we realize that for each combination of sets we either take a set or reject it. Thus in the case of two sets A and B, we can take A and reject B, take B and reject A, take both A and B, and finally reject both A and B, but this last case does not come into our Venn diagram; hence we have two options for each set, giving 2×2 or 2^2 options in all, but only $2^2 - 1$ spaces in the Venn diagram to represent the intersections of the two sets. Similarly for three sets we require $2^3 - 1$ or seven spaces, and for four sets we require $2^4 - 1$ or fifteen spaces, and we can check these totals with Figs. 11.1 and 11.2. In the general case of n sets, $2^n - 1$ spaces are required in the Venn diagram to represent the set intersections.

Looking again at Figs 11.1 and 11.2, both of these Venn diagrams have symmetry, but the one for four sets, Fig. 11.2, is symmetrical only about one axis and is not so pleasing in appearance as the one for three sets. One wonders if it is not possible to draw diagrams with curves (e.g. circles, ellipses) which would cover the situation.

The four-set diagram which has already been used gives a hint to a possible solution if ellipses are substituted for the closed U-shapes employed above. Having done this, can you develop a similar diagram for five sets and their intersections? The solutions to both these problems are given in the answers at the back of this book (*p.* 195), and it is interesting to note that they are the actual diagrams proposed by Venn himself. He did not give a diagram for six sets, but suggested the use of two five-set diagrams in conjunction and he made no statement about diagrams involving more than six sets.

The topic which we have been discussing so far is concerned with the division of a plane into spaces by drawing boundary lines. But if we are dealing with space, are we not considering geometry? Admittedly this is not the same kind of Euclidean geometry which has been taught for so long in our schools. Instead it is a geometry which is concerned with such thing as inside and outside, boundaries, surfaces and basic similarities between shapes; we call this study 'topology'. Its full development can lead to some quite abstract mathematics; in elementary books, such as this one, we only see some of the interesting practical applications.

Almost every book which gives a popular exposition of the newer mathematics includes something on topology and, in

I

turn, most chapters on topology mention the Königsberg bridges problem. This was first proposed and solved by Euler, and asks the reader to find a way over the seven bridges across the river at Königsberg (see Fig. 11.3) so that each bridge is crossed once and once only.

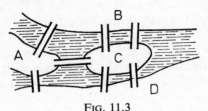

FIG. 11.3

In Which the Problem of the Königsberg Bridges is Exploded

The answer to Euler's problem is that it cannot be done. Rumour has it that a certain mathematician had been faced with this problem so often that he decided that not only could it not be done, but that it had been overdone. Thereupon he managed to smuggle some explosives into East Germany and blew up one of the bridges, hoping to end the problem for ever. The East German authorities have given no information but a report from unofficial sources indicates that he could not have been a very good mathematician since, given the condition that you must return to your starting point after you have crossed each of the bridges once and once only, the problem still cannot be solved for the six remaining bridges. Can you say which bridge the mathematician blew up? (*Solution follows.*)

The solution of the original Königsberg bridges problem is achieved by representing the map by an equivalent diagram

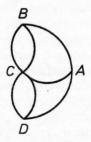

FIG. 11.4

in which the bridges are represented by lines and the land areas by points at which the lines meet (Fig. 11.4). Comparison with the bridge diagram will show how it represents the problem. For example, the line from *A* to *B* on this diagram represents the bridge from the island *A* to the bank of the river at *B*, the line from *A* to *D* represents the bridge from *A* to the bank of the river at *D*, and the line *AC* represents the bridge between the two islands *A* and *C*.

Consider the various possible combinations of lines which meet at a point, or—as they are here—routes which lead to a point, when you only use each route once.

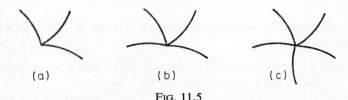

(a) (b) (c)

Fig. 11.5

If there are an even number of lines leading to any point, it is possible to come to that point and leave it each time (Fig. 11.5b above), but if there are an odd number of points, then on one occasion entry by a line gives no possible exit except by using a line already used (Fig. 11.5a and c above). So, in a diagram such as that above, a point with an odd number of lines leading into it may only be a starting point or a finishing point, and clearly one can only have two such points in any one diagram. If, however, one has to return to one's starting point, no odd-numbered points are permissible, and all the points must have an even number of lines leading into them.

Applying this rule to Fig. 11.4 we see that it is incapable of

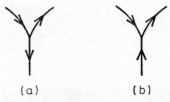

(a) (b)

Fig. 11.6 Points where an odd number of paths meet may only be a starting point (a) or a finishing point (b).

solution whether one states that one has to return to the starting point, or whether one does not make this stipulation, because there are four points each of which has an odd number of lines leading to it. As there can only be one starting point and one finishing point, only two such points are permissible if the problem is to be solved.

To deal with the case when one bridge is blown up, we have to remove a line from our simplified diagram so that only two points remain with an odd number of lines meeting there. Analysing the diagram we see that the removal of any one of the lines will give this condition, so that it does not matter which bridge is blown up, and it is impossible to state which it was.

European Tour A simplified map of Europe is given in Fig. 11.7.

    (i) Is it possible to travel by land through all the countries in turn, visiting each once and once only?

   (ii) Is it possible to tour through all the countries of Europe, crossing each frontier once and once only?

(Hint p. 185. Answer p. 195.)

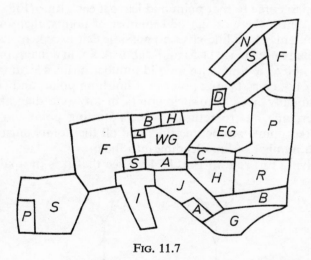

Fig. 11.7

A figure which can be drawn completely without taking pencil from paper and without covering any line twice is said to be unicursal; if, in addition, it is possible to return to the starting point with the final line, the figure is said to be closed. Thus Fig. 11.8 is unicursal since we may start at *A* and follow the path *ABCDEFGHAJ* without taking pencil from paper, but it is not closed since we cannot return to *A* without traversing the line *AJ* twice. How do you relate this to the fact that the figure has two points, *A* and *J*, with an odd number of lines meeting at each?

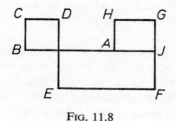

FIG. 11.8

Continuous Line State which of the following figures is unicursal, and if so, whether it is closed. (*Hint p. 185. Answer p. 195.*)

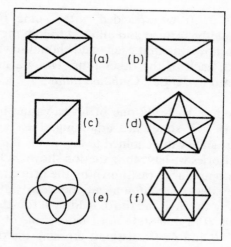

FIG. 11.9

Letter Problems Find which of the capital letters of the alphabet are unicursal. (*Answer p. 195.*)

All Around the House 1 If each room of a house has two

doors, and the house has two outer doors only, show that it is possible to enter the house by one outer door and leave it by the other, having passed through each room once and once only. (*Hint p.* 185.)

All Around the House 2 It is possible to enter a house by one door and leave it by another, having passed through every door in the house once only. What can be said of the number of doors in each room? (*Answer p.* 195.)

The Federation 1 The country of Wangola was split by tribal differences, so it was decided to separate the country into a number of federated states. In order that there should be the least risk of one state attacking another, it was decided that each of the four states into which the country was to be divided should have a common boundary with each of the other three. Thus, a state attacked by one of its neighbours could easily get support from the other two. Draw a map to show one possible solution to this division of the territory. (*Answer p.* 195.)

The Federation 2 It was decided, after some time, that a fifth state should be formed and allowed to join the Federation. The condition now was that each state should have a common boundary with each of the other four. Draw a map to show the division into five states. (*Solution follows.*)

This problem is basically one of investigating lines of communication. If two states are adjoining, then any point A, say, in the first state can be joined to any point B in the second. For three territories we have the situation shown in Fig. 11.10(a). If we draw in boundaries (as shown by the broken lines) we see that it is possible to communicate between any two states without passing through the third. Adding a fourth state means that any point D in that state must be capable of being joined directly to A, B and C, i.e. without crossing the other lines of communication. This can be done as shown in Fig. 11.10(b) so the four-state problem is capable of solution. Can you sketch in the appropriate state boundaries?

Bringing in a fifth state means that a fifth point, E, has to be added to Fig. 11.10(b) in such a way that it can be connected directly, and without crossing any other line of communication,

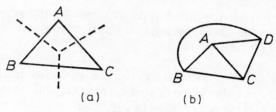

FIG. 11.10

to each of A, B, C, and D. In (b) the plane has already been divided into four regions—the three areas enclosed by the lines and the area around the outside of the figure. Now in whichever of these regions we place E, one of the other points must fall outside that region. For example, if we put E in the space

FIG. 11.11

bounded by the three lines AB, BC and CA, (Fig. 11.11) D lies outside that space. This means that we cannot join E to D without crossing one of the existing lines, and the same is true if we put E in any other of our spaces—each time there is one point which cannot be connected to E without crossing an existing line. So our second problem is incapable of solution. In other words, it is impossible to arrange five territories such that each has a joint boundary with each of the other four.

The foregoing example is a particular application of what has come to be known as the 'four-colour problem'. This states that only four colours are needed to distinguish between the different countries on a plane or spherical map, and is an interesting example of certain aspects of mathematical proof. We frequently speak, in mathematics, of a certain condition as being necessary and sufficient; note that the two are not synonymous. In the case of the four-colour problem, it can be shown that four colours are necessary to colour a map, but it has not been proved to date that four colours are sufficient.

Ring-ring Three circular metal rings are arranged so that they can not be separated until any one of the rings is cut, whereupon all three become separate. How are the rings arranged? (*Answer p.* 195.)

String-ring If the metal rings in the previous problem are replaced by circular loops of string, there are additional ways in which the same situation may be executed. In addition, some of these ways permit the problem to be extended to *n* rings, where *n* is any number. Can you find such a way? (*Answer p.* 195.)

These problems bring us on to the subject of knots in strings and ropes, which was one of the first branches of topology to be investigated mathematically. Nevertheless, there are still many properties of knots which are well-known but for which no mathematical proof exists. For example, it is known, but has not yet been proved, that it is impossible to tie a knot in a string, then to tie a second knot so that when the two are brought together they cancel each other out and the string becomes clear of knots.

Inside and Outside A certain Eastern potentate had an eye for a lovely girl, and also a reputation as a just and honorable ruler. Whenever he desired a girl he wished to show that he could win her by fair means and he would have her brought to the main hall of his palace before many of his followers. He would then instruct his guards to take a long rope and coil it on the floor in a double spiral as shown. The girl was told that she could stand in either of the central loops (*A* or *B* in Fig. 11.12). The

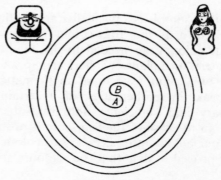

FIG. 11.12

potentate would then pick up the two ends of the rope in one hand, and draw them away. If the girl was within the rope that he was holding, she would be despatched to his harem; if she was not standing within the rope she would be free. Now the potentate happened to be a mathematician and *always* achieved his desire. How? (*Solution follows.*)

This puzzle shows that inside and outside are not always very clearly defined. Consider an area completely enclosed by a line such as Fig. 11.13. Clearly the inside is separated from the outside by the boundary line, and if we are inside the boundary and then cross it, we must be outside the curve. Now cross the boundary a second time and we must be inside the curve once more. We may repeat this process as often as we wish and obtain a general result. If we start outside a closed curve and cross the boundary n times, we are outside the curve if n is even, and inside it if n is odd. Starting from the inside of the curve, the reverse is the case.

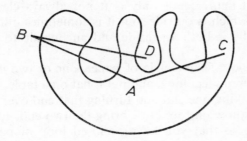

Fig. 11.13

(i) Starting at A inside the curve we cross the boundary an odd number of times to reach B outside the curve, and an even number of times to reach C within the curve.

(ii) Starting at B outside the curve, we cross the boundary an odd number of times to reach A inside, and an even number of times to teach D outside.

The method by which the potentate achieved his desire should now be clear. As he was outside the rope and he wished the girl to be inside it, he had to ensure that there was an odd number of lines of rope between himself and the girl when he picked it up. If she stood at A he would pick up the two strands of rope at the foot of the diagram, while if she stood at B he would pick them up at the top of the diagram.

The Closed Maze Mark any point somewhere near the centre of Fig. 11.14, and determine whether it is inside or outside the closed curve which constitutes the diagram.

FIG. 11.14

So far we have dealt mainly with two-dimensional figures. The subject becomes more complex when we deal with three-dimensional figures; not only is it not always clear which is inside and which is outside, but it is sometimes difficult to say if there really is both an inside and an outside to our figure!

Moebius Strips Take a strip of paper, one or two inches wide, and two or more feet long, and put it flat on a table. Give it half a twist by taking one end and turning that end over so that the underside is now on top. Now bring the two ends together and fasten them so that you have a closed loop of paper with a twist in it.

How many sides has the resulting figure? Check your answer by starting at one point on the surface and drawing a continuous pencil line.

How many edges has the figure? Check this by starting at one point on the edge and making a pencil line along the edge until you reach your starting point again.

What sort of figure would be obtained by cutting the closed strip down the centre? Try it with a pair of scissors. Was your answer correct?

Make another similar strip; what figure would you obtain by cutting the closed strip one-third of the width from one edge? Try it and see.

Instead of making a strip containing a half-twist, give the

strip a complete twist. That is, taking one end, turn it over until the upper side is uppermost again as shown in Fig. 11.15; now join the ends.

FIG. 11.15

Check the surfaces and the edges. What would be the result if this strip were cut down the centre? Can you discover any law connecting the number of twists (or half-twists) with the resulting cut figures?

Solid shapes may display various characteristics. Thus there are regular figures such as cubes, pyramids, spheres and so on; less regular ones which still display some ordered form, such as a tea cup, or a doughnut (or the mint with the hole!); those having no regular form at all, such as a child's modelling clay when he has finished playing with it and pushed it together in a lump; and those which may seem rather formless to some eyes yet contain artistic beauty, such as a Henry Moore sculpture, or even the ruins of an abbey. Is there any connection between these various types of figures in spite of the fact that they seem to differ so widely in shape and dimensions? In an earlier chapter we talked about finding patterns in mathematics and how it is part of the purpose of mathematics to bring order out of seeming confusion. It is interesting to see how the above-mentioned figures are classified.

We start by assuming that the solid objects which we are studying are flexible, so that they can be deformed from one shape to another. Thus a sphere may be pulled out into a sausage-shape, and this in turn may be bent round to form a horse-shoe shaped solid. Again a sphere may be flattened to form a thick disc, like a thick biscuit. However, we are not allowed to tear apart or join together. This clearly imposes a limitation; for example, we may take a sphere, stretch it to make it sausage-shaped, then bend it round to make it horse-shoe shaped, but we cannot bend it further to make the two ends meet and then join them together to form a doughnut shape (or as we usually call it, a torus). Hence a torus differs essentially from a sphere.

Deformations   Say which of the following can be transformed into any of the others by our imagined bending and squeezing:

(a) a sphere; (b) a dinner plate; (c) a tea-cup with a handle; (d) a cube; (e) an inner tube of a bicycle; (f) a soup bowl with two handles; (g) a piece of garden hose; (h) a jam jar; (j) the first of the Moebius strips made in the previous exercise. (*Answer p. 195.*)

What we have been considering is classification according to connectivity. This is best understood by imagining cuts in the solid in such a way that each cut starts at a particular point on the surface and returns to that point. If such a cut separates the surface of the solid into two or more distinct parts, we call the cut 'effective', since any point within the circle of the cut is separated from any point outside it. All closed curves drawn on the surface of a sphere are effective in this way, so we say a sphere has connectivity one.

The genus of a surface is defined as one less than the connectivity, so that the genus of a sphere is zero. The torus, however, illustrates better this type of classification (see Fig. 11.16).

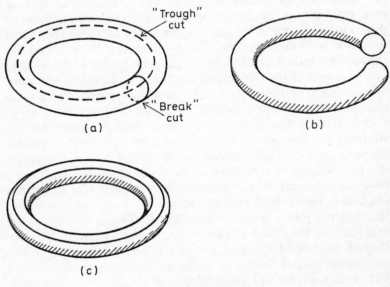

FIG. 11.16

We can make two types of cut on the surface of a torus: one which splits it open (making a kind of circular trough) and one which breaks the ring (making an open-ended pipe).

It is possible to make these two cuts successively on the torus, but any further cut will destroy the solid since it is impossible to make an additional cut without splitting the torus into two separate parts. So, in the case of a torus, the number of effective cuts is two, the connectivity is two, and the genus is one.

Additional methods of classification are needed, however, as can be seen by considering a gold-fish bowl. Like a complete sphere, this is of connectivity one, and yet we cannot transform a gold-fish bowl into a complete sphere, or a sphere into a gold-fish bowl, without tearing or welding (see Fig. 11.17).

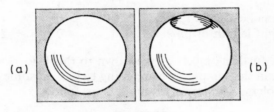

(a)   (b)

FIG. 11.17

The surface of the sphere is continuous, but the surface of the gold-fish bowl has an edge, so we have to consider what we mean by an edge, i.e. a termination of a surface. This is not quite the same as what is meant by the word 'edge' in normal geometry. For example, a cube has edges, yet a solid sphere may be transformed into a cube simply by stretching and remoulding. Thus an edge in the normal geometrical sense may mean a folding of a surface in a topological sense, whereas in topology an edge means a termination of a surface. Finally in our classification we must note the number of sides which our object has, although once more 'sides' has a somewhat different meaning topologically from the more usual meaning in geometry. Our first Moebius strip illustrates the importance of this, since no amount of deformation will transform this solid into a two-sided solid.

Transformation of solids, then, depends on the original solid having the same genus (or order of connectivity), the same number of edges, and the same number of sides as the solid into which it is to be transformed. Unless these agree, transformation of one into the other is not possible. Or—to put it another way—these three qualities remain invariant with change of the shape of a solid, provided always that the surface of the solid is not torn in the process.

It may be of interest to list one or two solids, with their corresponding constants, and to find from the list which solids may, or may not, be transformed into another.

	Genus	Edges	Sides
Solid sphere	0	0	1
Inflated ball	0	0	2
Jam jar	0	1	2
Torus	2	0	2
Tea cup*	2	1	2
Solid cube	0	0	1
Straight length of pipe	0	2	2

Add some further objects of your own to this list.

Finally, in this chapter, we are going to look at some 'shortest path' problems. A simple example is as follows:

Playground Circuit A class of children are playing in a school playground. The playground is rectangular in shape and they mark a point on the ground. The game is for each child to start from this point, touch, in turn, each of the four walls which surround the playground and return to the starting point, the winner being the one who does the circuit in the quickest time. Assuming that all the children can run equally fast, what would be the winner's path? (*Solution follows.*)

Clearly the winner's path will be the shortest one, and it is a well-known geometrical fact that the shortest distance between two points is a straight line. This hardly seems relevant in the present case since the path is bound to be a circuit, but the

* There may be some ambiguity in certain cases depending on whether we consider the thickness of the objects. In the table we consider the bowl of the cup to have no thickness, as we did earlier for the gold-fish bowl. The same applies to the pipe and the jam jar. Earlier in the text we said that a sphere could be flattened into a thick disc. Readers may care to reclassify in the above table if these objects are regarded as having a thickness.

problem may be rearranged to reduce it to finding a straight line between two points. The method is to represent each section of the child's path on a separate plan of the playground, successive plans being arranged with their corresponding sides adjoining; then the sections of the path form a straight line between the starting and finishing points, which are on different plans. This is shown in Fig. 11.18. The starting point is A and the finishing point is A', and a letter F is drawn on each plan, being reversed and/or inverted in each case to show the transformation of that particular plan. The straight line joining the

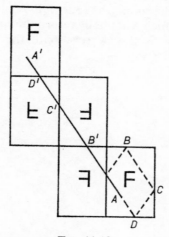

Fig. 11.18

two points A, A' represents the shortest distance, and the actual path may now be drawn on the first plan (dashed line). It will be seen that the circuit is a parallelogram.

The shortest path between two places on a sphere was mentioned briefly in Chapter 7; it is along a great circle passing through the two points. A great circle is the largest circular section of the sphere. Only one line of latitude (the Equator) is a great circle, so that (except at the Equator) the shortest distance from one place on the earth's surface to another lying on the same line of latitude does not follow that line of latitude. Thus a plane flying from London to Newfoundland by the shortest route would leave London not in a westerly direction but in a direction roughly north-westerly. The best way to illustrate this is to mark two points on a sphere (preferably a world globe, but a large ball would suffice). If a piece of string is now held in contact with one of the points and then pulled tightly to the other point, the great circle path may be seen.

Round the Cone What is the shortest distance from a point P on the surface of a cone, right round the cone's surface and back to P? (*Answer p.* 195.)

Spider Circuit In a room which is a perfect cube, a spider is at the middle point of one of the edges and wishes to walk on each of the surfaces of the room (the four walls, ceiling and floor) in a complete circuit which will finally return the spider to its starting point. What shape is the shortest path the spider can take? (*Answer p.* 196.)

CHAPTER 12

Interlude 3: Something More to Think About

First If you write four letters and address the four envelopes in which the letters are to be sent, clearly there is only one way in which all the letters may be placed in the correct envelopes. But how many ways are there of putting all the letters into the wrong envelopes? (*Answer p.* 196.)

Second Four volumes of an encyclopedia are arranged in order on a bookshelf. Each volume consists of four hundred pages. A bookworm starts at page one of volume one and eats it way in a straight line through to page 400 of the last volume. Ignoring the covers, through how many pages does the bookworm eat its way? (The answer is not 1600, by the way!) (*Answer p.* 196.)

Third Harry had a stall on the market and got his customers by selling fruit cheaply. One week he had a job lot of oranges, 180 in all, which he sold at 5 for 5p. Next week he got another 180 which were rather better and he sold them at 4 for 5p. The following week he got a further 180 of each kind, but instead of selling one lot at 5 for 5p. and the others at 4 for 5p., he put the two lots together, 360 of them, and priced them at 9 for 10p. Now he remembered getting 180 p. for his oranges the first week and 225p. the second week, so he expected 405p. for his sale on the third week, and was puzzled when he found that he had only got 400p. Can you explain why? (*Answer p.* 196.)

Fourth Write down the squares of the numbers from 1 to 9. Alongside these, write down the squares of the numbers from 11 to 19. Compare the last digits of each of the first set of squares with those of the second set. If necessary, write down the squares of the numbers 21 to 29. Then try to explain:

(a) Why the pattern of last digits is always the same;

K

(b) Why the first and last digits of each set are the same, the second and next to last, the third and third from the end, etc.

(Answer p. 196.)

Fifth The stations on an underground railway line are Parrock, Landon Crescent, Maplesfield, Duke Bar, Stenchford, Row's Fields, Dunston, Maple Grove, Larchfield High Street, and Dench. Now Len uses the underground daily to travel from his home in Maple Grove to his work in Parrock and he knows that the trains run at regular intervals along the line, turning round at each terminus and coming along the return route immediately; since the service is regular and frequent, he does not worry about going for any particular train but just goes to the station at Maple Grove and catches the next one that comes for Parrock. Over a period of time, however, he began to get the feeling that trains seemed to be running more frequently in one direction than in the other so he decided to do a survey, and each morning as he went to work he made a note of which train came first, one to Parrock or one to Dench. As he was going at no particular time, he knew he was taking a purely random sample, but his survey confirmed his suspicions, for he found that the number of occasions when the Dench train came first far exceeded the occasions on which the Parrock train came first. So he wrote a letter of complaint asking what happened to the trains at the Dench terminus. Have *you* any explanation? *(Answer p. 197.)*

Sixth In Chapter 9 I told the story of *The Fibs and the Trues*, and the other day I went to visit Pat who lives on the other side of town. I knew the street but not the number, though I rather thought it was twenty-seven. I was just hesitating outside that number when a teenager came out of the house next door. Was he a Fib or a True? I didn't know! But what question should I have asked him so as to be quite certain whether or not Pat lived at number twenty-seven? *(Answer p. 197.)*

CHAPTER 13

Mathematics Miscellaneous

In this chapter some problems which we shall meet are concerned with relationships between numbers. Most of these relationships are simple ones, but we are being asked to recognize them in new situations. The first type of puzzle consists of coding or suppression of numbers in calculations involving the four basic rules—addition, subtraction, multiplication and division. In some of these problems letters replace the digits and we have to find the code—which digit is represented by which letter? In others some of the digits are replaced by a dash or a cross.

It would tend to spoil the fun of this type of problem if we gave too many hints on how to solve them; it is almost possible to make a list of rules which would enable anyone to solve the problems by simple application of them, but there are probably no other types of puzzle where almost everyone already has the basic knowledge, and so can develop their own methods of solution. As a simple example, suppose we know that three digits are represented by A, B and C, and that a multiplication may be written as

$$\begin{array}{r} \mathbf{AB} \\ \mathbf{C} \\ \hline \mathbf{AB} \end{array}$$

then clearly C must be 1. This gives us no clue to A and B, of course, which might be anything, and this is one of the snags of this type of problem. Quite often it is possible to have more than one solution to the puzzle. An example of this was the *Work this out* problem of Chapter 1; one solution was given immediately following the problem, another two are given in the answers at the back of the book, and still others are possible. At the other extreme it is possible to pose a problem which at first glance appears to be capable of solution (i.e. the number of digits at each stage may be correct), but proves to have no solution at all. Between these extremes there are those puzzles which have only one solution, or—as we say—the solution is unique.

The property of 1 in a multiplication problem, as shown above, is paralleled by 0 in an addition or subtraction problem. Multiply by one and the result is the same; add or subtract 0 and the result is the same. These are simple instances of the number facts which we apply to these puzzles, but more complex ones are obviously called for. A useful way of starting a problem is to make a table containing the digits 0 to 9 and then insert the letter opposite to its appropriate number as each is found. You are now invited to

Face Up to It Solve the following addition sum

$$
\begin{array}{c}
\text{H A N D} \\
\text{N O S E} \\
\hline
\text{E Y E S}
\end{array}
$$

(Solution follows.)

The first point to note is that H + N gives a single digit, and we know that a number may not commence with a 0, so neither H nor N represent 0. Next H and N give an answer E and so do N and S, which means that there must be a carry over of one either from the first to the second column or from the third to the fourth (counting the columns from the right). In other words there is a difference of one between H and S. From the first E of the word EYES we can say that E does not represent 0, 1 or 2. (Why?) With these facts, and noting that N and S appear twice and E three times, we may try figures in various positions. Proving that any solution is unique is not always an easy matter, since the obvious method of trying to find a second solution, or even third or fourth, may be limited by one's ability rather than by the problem! A solution of the above problem is

$$
\begin{array}{c}
2 \ 4 \ 6 \ 3 \\
6 \ 5 \ 1 \ 8 \\
\hline
8 \ 9 \ 8 \ 1
\end{array}
$$

Is this a unique solution?

The next puzzle is an example of the suppressed number type

of problem. Numbers are to be inserted in the spaces, one digit to each dash.

Seven and Eight

$$- -) - - 7 - - (8 - - -$$
$$- -$$
$$\overline{}$$
$$- -$$
$$- -$$
$$\overline{}$$
$$- - -$$
$$- - -$$
$$\overline{}$$

(Solution follows.)

In this division the first partial product consists of 8 multiplied by a two-figure number and the result is also two-figured. This means that the divisor is either 10, 11 or 12. The last partial product is a three-figure number, and this tells us that the divisor must be multiplied by a number greater than 8; this can only be 9, so the divisor must be 12. We now start at the beginning again; the first partial product is 96, and after subtraction we have a number, 7, but this must give a remainder in turn of only 1 (the first figure of 108). So the second partial product must be 36 and we may now write out the complete problem as follows:

$$12) \ 99708 \ (\ 8309$$
$$96$$
$$\overline{}$$
$$37$$
$$36$$
$$\overline{}$$
$$108$$
$$108$$
$$\overline{}$$

In the example above, the solution is quite clearly unique, and this is also true of the following puzzle which is of ancient origin.

Long Division

```
- - ) - - - - - - ( - - -
        - 0 - -
        ─────────
        - - - -
        - 5 0 -
        ─────────
          - - -
          - 4 -
```

(*Hint p.* 185. *Answer p.* 197.)

Multiplication 1 The letters in the following multiplication problem each represent a digit.

```
        H O T
        H O T
      ─────────
      L A S T
    T A O P
  A M M P
  ─────────────
  A H Z S S T
```

(*Hint p.* 186. *Answer p.* 198.)

Multiplication 2

```
      M B C F
        A G H
    ───────────
      B J G H
  M M J F J
  C J B K
  ─────────────
  B M C A A H
```

(*Hint p.* 186. *Answer p.* 198.)

A Lot of Subtraction The matter of unique solutions and multiple solutions has been discussed above. The following is a subtraction problem with twenty-four solutions. Can you find all of them?

```
B D F H A
C E G I J
───────────
A A A A
```

(*Hint p.* 186. *Answer p.* 198.)

Letter Division   The following is a long division problem.

```
B E ) F D B E ( C B A
      F E
      ───
      D B
      D E
      ───
        B E
        B E
        ───
```

(*Hint p.* 186. *Answer p.* 198.)

On All Fours

```
– 4 – ) – – – – – ( – – –
        – – 4
        ─────
        – – –
        – – 4
        ─────
        4 – –
        4 – –
        ─────
```

All the fours are given. (*Answer p.* 198.)

Most of the puzzles in this book involve mathematical principles which are important in other applications, such as various branches of science, technology and commerce. The mathematics is the same, the only difference being the use made of it. Some parts of mathematics, however, are interesting in themselves but seem to have little practical application. One such topic is magic squares, although at one time these were regarded as charms or talismans which possessed special properties.

In Chapter 3 we had a puzzle, *Digital Sum 3*, which was

simply a well-known problem in a slight disguise. Basically the problem was to arrange the figures 1 to 9 in a square so that the sum of the figures in any row, column or diagonal is the same. There is only one basic solution to this problem, as shown in Fig. 13.1.

4	3	8
9	5	1
2	7	6

FIG. 13.1

It is possible, of course, to interchange certain rows or diagonals as, for example, in Fig. 13.2, but careful examination

8	3	4
1	5	9
6	7	2

4	9	2
3	5	7
8	1	6

FIG. 13.2

will show that these are really the same combinations. In fact, a total of eight variations of the basic square are possible. Can you write them down?

The 3 × 3 magic square is said to be of the third order. Lower order magic squares are not possible, but squares of the fourth

(a)

16	3	2	13
5	10	11	8
9	6	7	12
4	15	14	1

(b)

10	2	9
6	7	8
5	12	4

FIG. 13.3 (a) A fourth order magic square. (b) A third order magic square with numbers other than the first nine digits.

and higher orders may be constructed. In these cases, though, we are no longer restricted to one basic pattern. It is possible to construct magic squares with different groups of numbers, but we shall only consider here those made from the first n integers starting at 1 and counting upwards, i.e. a third-order square made from the numbers 1 to 9, a fourth-order square made from the numbers 1 to 16, etc.

To construct a magic square it is clearly an advantage to know the sum in any diagonal, row or column. So we pose the problem:

Magic Square Sum   Find the sum of the numbers in any row, column or diagonal of a magic square of order n, starting with 1 and containing the first n^2 digits. (*Solution follows.*)

If the square is of order n, it contains n^2 numbers. Consider the rows containing the number n^2. One of these rows must contain the sum made by combining n^2 and 1, since this is essential to obtain a minimum total. But remembering the earlier work in this book on arithmetic series, we may make many other combinations with the same total $n^2 + 1$. Thus if n is 4 and n^2 is 16, we have

$$\begin{array}{cc} 16 & 1 \\ 15 & 2 \\ 14 & 3 \\ 13 & 4 \\ 12 & 5 \end{array}$$

and so on.

(What happens if n is odd, so that we have $\frac{1}{2}(n-1)$ pairs and an odd number left over? What is this odd number?)

The pattern above shows clearly the combinations which appear in the various rows. If there are n numbers in a row, there must be $\frac{1}{2}n$ totals, each of $n^2 + 1$. So that the total of any row or column must be $\frac{1}{2}n(n^2 + 1)$. In the case of odd values of n, the formula still applies because of the odd term referred to in the parenthetical note above. Also in this case, the middle term of the series is also the middle term of the square. Thus with a square of order 3, the term occupying the centre square is 5, with one of order 5, the term is 13, and generally if the order is n, the central term is $\frac{1}{2}(n^2 + 1)$. This is one of the reasons why magic squares of odd order are more easy to construct than those of an even order.

Several interesting methods have been devised for constructing magic squares of odd and even orders, and the reader interested in these is referred to the references at the end of this book.

A Fourth Order Magic Square   Construct a magic square from the numbers 1 to 16, so that the number 1 falls in the top left-hand square. (*Answer p.* 198.)

Somewhat similar to magic squares, but different in construction are Latin squares. Surprisingly, these have a practical application, being important in experimental crop testing and rotation, etc. They are constructed by taking a square with n^2 spaces in it, and arranging the numbers 1 to n in each of the rows and each of the columns, as for example in Fig. 13.4.

1	2	3
2	3	1
3	1	2

FIG. 13.4

It will be noticed that each of the numbers appears once and once only in each row and in each column. If, in addition, the diagonals also satisfy this condition, the Latin square is said to be diagonal.

Diagonal Latin Squares   Construct a diagonal Latin square using the numbers 1 to 4. Is it possible to do the same for the numbers 1 to 3? (*Answer p.* 198.)

We now embark on an excursion into modular arithmetic, a subject important in the theory of numbers, which will eventually bring us back to Latin squares. In modular arithmetic we are concerned only with remainders when numbers are divided by some base. Thus 12 expressed to the base 8 in this notation would be known as 4 (modulo 8), because when 12 is divided by 8 it leaves a remainder of 4. Similarly 12 would be equivalent to 2 (modulo 10), since 12 divided by 10 leaves a remainder of 2. 12 would also be equivalent to 2 (modulo 5) since the remainder is also 2 when 12 is divided by 5. Notice that 12 is equivalent to 0 (modulo 6) since 6 divided 12 exactly.

Just as in ordinary arithmetic we can make composition tables for multiplication and addition (see *Mixed Product*, Chapter 1), so we can make composition tables in modular arithmetic. The example below is the multiplication table for modulo 6. We arrange our numbers down the side and along the top or bottom, and where a row or column intersect we write the appropriate product. Thus 4×4 gives 4 (modulo 6).

5	0	5	4	3	2	1
4	0	4	2	0	4	2
3	0	3	0	3	0	3
2	0	2	4	0	2	4
1	0	1	2	3	4	5
0	0	0	0	0	0	0
×	0	1	2	3	4	5

Notice that the digits are 0, 1, 2, 3, 4, 5, i.e. six elements including 0 and not including 6, and this applies generally. For example, in any system to modulo 4, the only elements would be 0, 1, 2 and 3. Study the above table, noting particularly any patterns in the numbers, symmetry, etc.

Modulo 5 Composition   Complete the arrays below in order to make up addition and multiplication tables for modulo 5. (*Answer p.* 199.)

4	4				
3	3				
2	2				
1	1				
0	0	1	2	3	4
+	0	1	2	3	4

4	0	4			
3	0	3			
2	0	2			
1	0	1	2	3	4
0	0	0	0	0	0
×	0	1	2	3	4

In the above composition tables the bottom rows and the left-hand columns progress in ones, or in other words they are successive digits, but this need not always be the case.

Modulo 4 Addition Draw up an addition composition table modulo 4, with the digits 0, 1, 2, 3 in the column and multiples of three (in modulo 4) along the row whose elements are being added. What do you notice? (*Solution follows.*)

The completed composition table is given below. Notice that the pattern formed is a Latin square. Compare with the addition table for modulo 5.

3	2	1	0	3
2	1	0	3	2
1	0	3	2	1
0	3	2	1	0
+	3	2	1	0

Modulo 5 Addition Repeat the above example in modulo 5 with the left-hand column progressing in ones, and the bottom row multiples of 2. What do you obtain this time? (*Answer p. 199.*)

Latin squares which contain all the elements once and once only in each of the diagonals as well as in the rows and columns are called diagonal Latin squares. Combination of two such squares which satisfy certain conditions (as, for example, in Fig. 13.5) may be used to lay out the face cards of a pack of cards so that no suit appears more than once in each column, row and diagonal, and similarly for each face value.

11	22	33	44
34	43	12	21
42	31	24	13
23	14	41	32

FIG. 13.5

Try giving the values of 1, 2, 3 and 4 to each suit and to each face value and check the layout of the cards. There are 72 basic solutions to this problem which, by reflection, etc., may be increased to 504 solutions in all.

The three problems which follow are extensions of this work.

The Magic Cube   Arrange the numbers 1 to 27 in three magic squares so that when they are placed one above the other they form a magic cube, i.e. columns and rows in two horizontal directions are magic. If possible try to achieve the magic property in the diagonals of the cube. (*Answer p.* 199.)

Elements of Modulo 8   Make up a multiplication composition table (modulo 8) for the elements 1, 3, 5, 7. Comment on your result. (*Answer p.* 200.)

A Switching Table   There are two electrical switches, numbered 1 and 2, and you are given a certain code of operation of these switches.

A means do nothing.
B means turn switch 1
C means turn switch 2
D means switch both 1 and 2.

Note that we are not concerned with whether the switch is 'on' or 'off', the instruction is simply to turn the switch the other way, i.e. off, if it is 'on', and on, if it is 'off'. Notice also that we may combine these operations to form one of the other operations. Thus *B* followed by *C* is equal to *D*, and *C* followed by *C* is equal to *A*. Now make up a composition table by putting *A*, *B*, *C* and *D* down the side and along the bottom in the same way that you did with 1, 2, 3, 4 in *Elements of Modulo 8*. What do you notice? (*Answer p.* 200.)

Finally, here are some puzzles and games using apparatus which may be made quite easily with cardboard or wood. Solutions of these are not given in the answers deliberately; you just have to puzzle them out!

Bisected Tetrahedron   Take some thin card—two postcards will do—and make two copies of Fig. 13.6 on the card. Cut round the outside of each figure, and fold it along each of the four inner lines in the same direction. Stick adjacent edges together, either with adhesive tape or by cutting the figures out with flaps as indicated by the dotted lines in the diagram, and sticking adjacent edges on to the flaps. You now have two identical solids. Can you fit them together to form a tetrahedron?

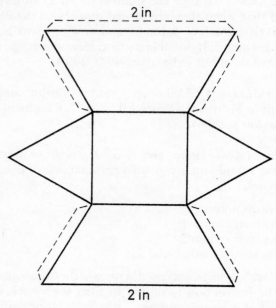

2 in

2 in

FIG. 13.6. All lines are 1 inch long except the two marked 2 inches.

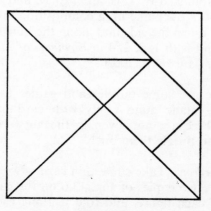

FIG. 13.7

Tangrams   Of the many dissection puzzles, tangrams is probably the oldest and best-known. It is easily constructed out of card. Take a square piece of card and mark it as shown in Fig. 13.7, then cut along all the lines. The problem is to make up certain figures using all the pieces of the puzzle; the pieces must not overlap. The simplest start is to mix up the pieces and then try to re-arrange them in the form of the square.

Next arrange the pieces to form a rectangle. Then a right-angled triangle, then a parallelogram. Many other figures may be constructed, often representing figures or objects. A few are shown in Fig. 13.8.

FIG. 13.8

Soma Cubes   Soma cubes may be regarded as a three-dimensional form of tangrams. They are of much more recent origin and were thought up by a Danish puzzle maker, Piet Hein. They consist of seven figures made up of cubes, six of which are made from four cubes each and cover all the possible ways of fastening four cubes together apart from a straight line. The seventh consists of three cubes (which may be put together in one way only). To make a set of these it is easiest to obtain a long stick of square cross-section (¾ inch square is a good size), and cut off the required lengths from this, then glue them together as shown in Fig. 13.9.

Having made the set, first fit them together so that they form

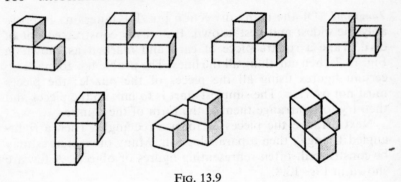

FIG. 13.9

a cube (there are two possible ways). Then find what other shapes may be formed by fitting them together.

Line Exchange Peg board may be purchased from most handy-man's stores, and a simple game may be constructed by taking a strip of this containing eleven holes. You now require some coloured pegs, five of one colour and five of another, which are then placed in the strip as shown in Fig. 13.10.

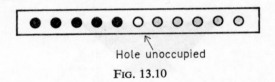

Hole unoccupied

FIG. 13.10

The game is to exchange the pegs of one colour with those of the other colour by moving one peg at a time either into an adjacent empty hole or to an empty hole which lies immediately behind an adjacent peg of the other colour. Backward moves are not allowed, that is, in the diagram above, black pegs may be moved only to the right, and grey ones only to the left. The real object of the game is to be able to perform it in the smallest possible number of moves.

Square Exchange This problem is a two-dimensional version of *Line Exchange*. Again the baseboard may be constructed from peg board and consists of two 3 × 3 squares with a common hole in the middle. This middle hole is initially without a peg, and eight pegs of one colour occupy the rest of one square, while eight of another colour are opposite, as shown in Fig. 13.11. The rules of the game are exactly the same as before. A peg may be moved forwards only, either into an adjacent empty hole or over a peg of a different colour to an empty hole immediately behind that peg. Diagonal moves are not allowed.

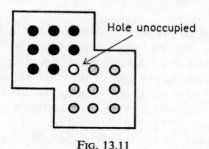

Hole unoccupied

FIG. 13.11

The object, once again, is to interchange the two colours in the minimum number of moves.

CHAPTER 14

Hints for Solutions

Note The solutions outlined below must not be regarded as the only acceptable method in each case. There is frequently more than one way to solve a mathematical problem, and it is often a valuable exercise to attempt to solve the same problem by a number of different methods; in some of the following cases more than one method is given. The methods which are given are chosen either for ease of explanation and understanding or to illustrate some particular technique. If readers find other methods, this is all to the good!

CHAPTER 1

Suspects Frazer's statement gives the clue. If his first statement is a lie and his second the truth, he is saying that he did it, but he doesn't know who did. This is impossible, so he didn't do it. Now proceed to consider Thomson's statements and then those of James. (Also see Chapter 9 of the text for another method.)

Pupil's Choice Make a table as shown below and fill in with a cross the combinations which are not allowed (N.B. it is not necessary to complete the table above the dashed line.)

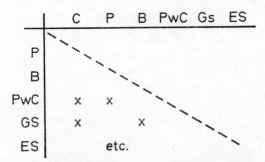

Board Meeting Match those who attended two meetings with those who attended one, and similarly for three attendances and none. (Alternatively use a Venn diagram, see Chapter 6.)

CHAPTER 2

Advanced Warning Use the binary system.

The Present Call written number *abc*, where *a*, *b* and *c* represent digits. In Xanto's five-based counting the number would be $c + 5b + 25a$. Zimba's reversed number would be written *cba* or would equal $a + 6b + 36c$ in his base-six counting. These are equal, so

$$c + 5b + 25a = a + 6b + 36c$$

i.e. $b = 24a - 35c$, where *a*, *b* and *c* are different integers less than five.

Bases and Powers

$$11^2 = 121$$
$$11^3 = 1331$$
$$11^4 = 14641$$

and so on.

Elevenses In base ten let number be *abcd*, whose actual value is $1000a + 100b + 10c + d$. Reversing gives the value $1000d + 100c + 10b + a$, and addition of these gives $1001a + 110b + 110c + 1001d$, which is divisible by eleven. Subtraction gives $999a + 90b - 90c - 999d$ and is always divisible by nine.

In other number bases, let base be *n*; then numbers are $n^3a + n^2b + nc + d$ and $n^3d + n^2c + nb + a$. On addition we have

$$(n^3 + 1)a + (n^2 + n)b + (n^2 + n) + (n^3 + 1)d$$

Since $n + 1$ is a factor of each term, the whole is divisible by $n + 1$. Try different values of *n* in the above.

CHAPTER 3

Digital Sum 1 The central number must be 7 and the sum of the numbers in any one line must be 21. (Why?)

Digital Sum 2 Since every number is used twice, the sum of the totals of the six lines must be twice the sum of the numbers from 1 to 12, i.e. 156. As there are six lines, the sum of the numbers in each line must be 26.

Digital Sum 3 The centre number is used eight times and all the others three times each. The centre number of the system must be the centre number of the pattern.

Odd Sums   Pattern is

$$1 = 1^3$$
$$3+5 = 2^3$$
$$7+9+11 = 3^3$$
$$13+15+17+19 = 4^3$$

and so on

The fourth row may be written as

$$4 \times 12 + (1+3+5+7) = 4^3$$

and similarly for the others. For the *n*th row we have

$$n^2(n-1)+[1+3+5+ \ldots +(2n-1)]$$
$$= n^2(n-1)+\tfrac{1}{2}n[2+2(n-1)]$$
$$= n^2(n-1)+n+n(n-1)$$
$$= n^3$$

Multi-Shaped Tesselations

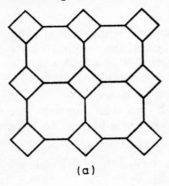

(a)

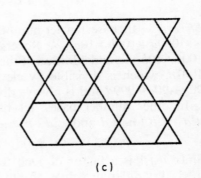

(c)

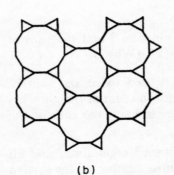

(b)

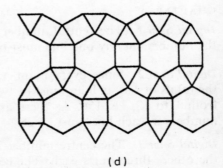

(d)

Fig. 14.1

CHAPTER 4

John and Mary Let x be John's present age. Mary's age when she was five years older than her age when she was a quarter of John's present age was $\frac{1}{4}x + 5$, Hence

$$\tfrac{1}{2}x = \tfrac{1}{4}x + 5$$

Now solve.

Partners Let x be Benson's original investment. Ames must receive £$\frac{1}{2}x$ from Charles' money to make investments equal, and the rest of Charles' money is divided equally. So Benson gets $\frac{1}{2}(2500 - \frac{1}{2}x)$ and this from his initial investment equals 2500 giving $x - (1250 - \frac{1}{4}x) = 2500$.

Joan and Bill Let x be Bill's age and y be Joan's age. When Bill was Joan's present age (i.e. y), Joan was $y - (x - y)$. But Bill's present age is twice this. Hence

$$x = 2y - 2(x - y)$$
$$3x = 4y \qquad (1)$$

Also when Joan is x years old, Bill will be $x + (x - y)$, so

$$x + x + (x - y) = 63$$
$$3x - y = 63 \qquad (2)$$

Solve (2) by substitution from (1).

Trenching Since the trench is of constant width, the amount dug out is proportional to the area shown in Fig. 14.2.

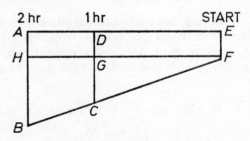

FIG. 14.2

Or, since the volumes in the two hours are the same, the area of the trapezium $ABCD$ is equal to the area of the trapezium

DCFE (The area of a trapezium is half the sum of the parallel sides times the distance between them). Now if *DE* is *l*, *AD* is $\frac{1}{2}l$,

$$\text{area } DCFE = \tfrac{1}{2}(EF + DC) \times l$$
$$\text{area } ABCD = \tfrac{1}{2}(DC + AB) \times l$$

Since these are equal

$$\tfrac{1}{2}l(EF + DC) = \tfrac{1}{4}l(DC + AB)$$
$$2(EF + DC) = (DC + AB)$$
$$DC = AB - 2EF \qquad (1)$$

Now draw a line *HGF* parallel to *ADE* as shown. From similar triangles *FHB*, *FGC*,

$$\frac{FG}{FH} = \frac{GC}{HB}$$

or

$$\frac{l}{1\tfrac{1}{2}l} = \frac{GC}{HB} \qquad (2)$$

Whence $GC = \tfrac{2}{3} HB$. Equation (1) may be written

$$(DG + GC) = (AH + HB) - 2\,EF$$

Use equation (2) to substitute for *GC* and obtain

$$EF = \tfrac{1}{6} HB$$

and the final answer may be obtained.

[For enthusiasts: what happens if he continues to dig in the same manner for another hour?]

Fly Catcher   The easy method is to calculate how long the fly is flying. Since the trains are coming together at 80 m.p.h. and are 160 miles apart at the start, the total time before the crash is two hours and during this time the fly is flying at 60 m.p.h.

Birthday Party   If sons' ages are *x*, *y* and *z*,

$$x + y + z = 59 \qquad (1)$$
$$x = 2y \qquad (2)$$

On his eightieth birthday, ages of the two whose present ages are *x* and *y* will be $x + 20$ and $y + 20$, and

$$x + 20 + y + 20 = 79$$
$$x + y = 39 \qquad (3)$$

Substitute (3) in (1) to obtain z, and in (2) to find x and y.

Procession   Speed of procession is 2 m.p.h. Let marshall's speed be x m.p.h.

Speed at which he passes procession in going to back is $x + 2$ m.p.h. and he passes $1\frac{1}{2}$ miles of it, so time taken is $1\frac{1}{2}/(x+2)$ h.

Similarly on his return he passes $\frac{3}{4}$ miles of procession at $x - 2$ m.p.h., so his time is $\frac{3}{4}/(x-2)$. But

$$\frac{1\frac{1}{2}}{x+2} + \frac{\frac{3}{4}}{x-2} = \frac{3}{8}$$

This gives

$$4(x-2) + 2(x+2) = (x+2)(x-2)$$
$$x^2 - 6x = 0$$

Alternatively it is possible to do the problem without algebra as follows. The marshall is away for $\frac{3}{8}$ h and clearly takes as long to reach the end of the procession as he does to return, so each time is $\frac{3}{16}$ h. In $\frac{3}{16}$ h the procession moves $\frac{3}{8}$ mile, so when he reaches the end, the procession has moved this distance and he has travelled $1\frac{1}{2} - \frac{3}{8} = 1\frac{1}{8}$ mile. As he travels $1\frac{1}{8}$ mile in $\frac{3}{16}$ h, his speed may be found.

[*Extension of the problem:* if the marshall rides past me and, at the same speed, goes to the head of the procession, then returns immediately, where is the end of the procession when he gets back to my position?]

Worth a Note   Since the value of the £5 notes equals the value of the £10 notes, if there are x £10 notes, there are $2x$ £5 notes. Let there be z £1 notes.

Since the statement is that $2z$ must be the least value to 'have more' £1 notes than £5 and £10 notes combined, equality must be obtained by subtracting one, i.e.

$$2z - 1 = x + 2x$$
$$3x = 2z - 1 \qquad (1)$$

Also

$$5x = \tfrac{3}{4}(z + 5x)$$
$$x = 3z/5 \qquad\qquad (2)$$

Substitute in (1).

Spot the Numbers The property of the roots of an equation may be proved as follows:

Suppose the equation $x^3 + px^2 + qx + r = 0$ to have solutions x_1, x_2, and x_3. Then it follows that the given equation could be written in factorized form

$$(x - x_1)(x - x_2)(x - x_3) = 0$$

Multiplying out, this last equation becomes

$$x^3 - (x_1 + x_2 + x_3)x^2 + (x_1 x_2 + x_2 x_3 + x_3 x_1)x - x_1 x_2 x_3 = 0$$

If this is equivalent to the initial equation, the following relationships must be true:

$$x_1 + x_2 + x_3 = -p$$
$$x_1 x_2 + x_2 x_3 + x_3 x_1 = q$$
$$x_1 x_2 x_3 = -r$$

which is the result used in the text.

The Narrow Bridge Let speed at which man runs be x m.p.h., and length of bridge be l miles.

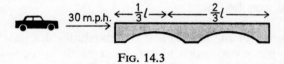

FIG. 14.3

Time taken for man to reach end of bridge nearer to car is

$$\frac{\tfrac{1}{3}l}{x} = \frac{l}{3x} \text{ h}$$

Time taken for man to reach other end is $2l/3x$ h. But the difference between these is the time taken for the car to travel at 30 m.p.h. for the length l of the bridge. This time is $l/30$ h. Hence

$$\frac{l}{3x} = \frac{l}{30}$$
$$x = 10 \text{ m.p.h.}$$

Right-angled Triangles 2 Let one triangle have sides x_1, y_1, z_1, and the other triangle have sides x_2, y_2, z_2.

$$x_1{}^2 + y_1{}^2 = x_2{}^2 + y_2{}^2$$

Applying the same notation to the equations given in the text

$$(a_1{}^2 - b_1{}^2)^2 + 4a_1{}^2 b_1{}^2 = (a_2{}^2 - b_2{}^2)^2 + 4a_2{}^2 b_2{}^2$$
$$(a_1{}^2 + b_1{}^2)^2 = (a_2{}^2 + b_2{}^2)^2$$

Thus we require the squares of two numbers (one odd, one even) to equal the squares of two other numbers, also one odd, one even. The lowest such combination of numbers is 1 and 8, and 4 and 7, since $1^2 + 8^2 = 4^2 + 7^2 = 65$. Substitution in the equations in the text gives the corresponding values of x_1, y_1, x_2, y_2.

Baling Out Let x = number of bales originally.

At the first corner, the number falling off is $\frac{1}{3}x + \frac{1}{3}$ leaving $\frac{2}{3}x - \frac{1}{3}$ still on the trailer. After the second corner, the number remaining is

$$\tfrac{2}{3}\left(\tfrac{2}{3}x - \tfrac{1}{3}\right) - \tfrac{1}{3}$$

Repeat this pattern and obtain as the number remaining after the fourth corner

$$\tfrac{2}{3}\left[\tfrac{2}{3}\left\{\tfrac{2}{3}\left(\tfrac{2}{3}x - \tfrac{1}{3}\right) - \tfrac{1}{3}\right\} - \tfrac{1}{3}\right] - \tfrac{1}{3}$$
$$= \frac{16}{81}x - \frac{65}{81} = n, \text{ say}$$

Hence $81n = 16x - 65$ where x and n have both to be integral. Since

$$x = \frac{81n + 65}{16} = 5n + 4 + \frac{n+1}{16}$$

for x to be integral, $n+1$ is a multiple of 16. Put $n = 15$ and $x = 80$. The requirement that at each corner only whole bales drop off remains to be checked.

Sum and Product This is an example where the method of trial and error seems to yield a result as quickly as any systematic method. To do the problem by algebra, suppose the number to be $10a + b$, where a and b are the digits. The relationship is then

$$10a + b = kab$$

(where k is a whole number). This may be written

$$b = a(kb - 10)$$

and substitutions made.

Small Change Denote the number of 5p. pieces by a and the number of 2p. pieces by b. We obtain the equation

$$5a + 2b = 80$$

Hence

$$a = 16 - 2b/5$$

Substitution will give the values, but this is not asked for; b must be a multiple of five and there are nine values which it can take (including $b = 0$).

Food Mixer Suppose he takes x ounces of A and y ounces of B.

$$12x + 5y \geqslant 120$$
$$4x + 4y \geqslant 45$$
$$x \geqslant 0$$
$$y \geqslant 0$$

Cost $= x + \frac{1}{2}y$

The graph is shown in Fig. 14.4, the initial profit line being 15p. or $x + \frac{1}{2}y = 15$

Oil Production Let l tons be the daily output of light oil and h tons be daily output of heavy oil.

$$h + l \leqslant 6000$$
$$l \leqslant 3000 + \tfrac{1}{2}h$$
$$h \leqslant 4250$$
$$l \geqslant 0$$
$$h \geqslant 0$$

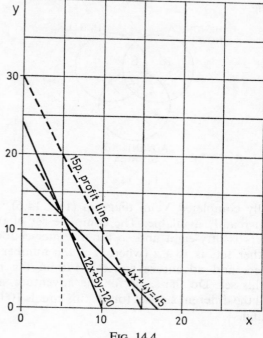

FIG. 14.4

CHAPTER 6

Motorist's Choice Since $\frac{1}{3}$ did not use the petrol and $\frac{1}{5}$ did not use either the petrol or oil, $\frac{1}{3} - \frac{1}{5}$, i.e. 2/15 use Zoom oil. Similarly it is found that 7/40 use Zoom petrol. The fraction who use both oil and petrol

$$1 - \left[\frac{1}{5} + \frac{2}{15} + \frac{7}{40}\right] = \frac{59}{120}$$

But this fraction is represented by what number of motorists? The problem may be illustrated or worked alternatively by a Venn diagram (Fig. 14.5).

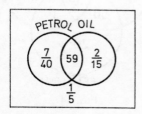

FIG. 14.5

Light Reading

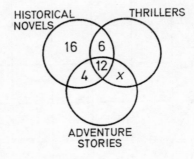

FIG. 14.6

The partially completed Venn diagram (Fig. 14.6) shows the information readily available. The complete set of those who liked thrillers is fifty-eight and so far the intersections of this with the other sets is $18 + x$ (where x is the number who like both thrillers and adventure stories). Fill in the remaining spaces in this set. Do the same for the adventure stories set, then add all the different sub-set totals; this equals 100, enabling x to be found.

The College Students A Venn diagram is as shown in Fig. 14.7, x and y representing the sub-sets shown. Equating totals of sub-sets to 100 gives $x + y = 43$. Note that it is not necessary to find the separate values of x and y in order to obtain the required result.

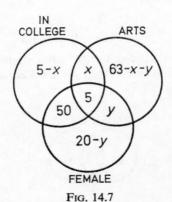

FIG. 14.7

Market Survey Draw a Venn diagram and find the totals. Compare with the survey totals as given.

Language Laboratory

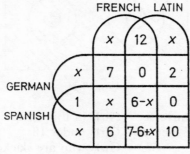

FIG. 14.8

Through Trains

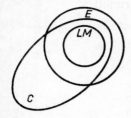

FIG. 14.9

E = Set of electric trains
LM = London to Manchester trains
C = Trains which pass through Crewe

Top People

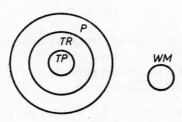

FIG. 14.10

TP = Set of top people
TR = *Times* readers
P = People who pay 5p. for a daily paper
WM = Working men

Office Routine

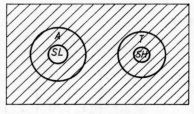

Fig. 14.11

Universe = Girls in this office
Set *A* = Girls under 18
SL = Girls who are slackers
T = Typists
SH = Girls who do shorthand
Shaded area = Girls who are brunettes

Animal Crackers

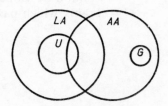

Fig. 14.12

LA = Set of legendary animals
U = Unicorns
AA = Absurd animals
G = Giraffes

Shade the area where the set lies which I am known not to like.

Committee Members

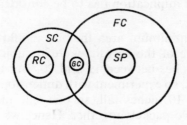

Fig. 14.13

SC = Social Committee
RC = Refreshment Committee
GC = General Committee
SP = Sports Committee
FC = Finance Committee

CHAPTER 7

Navigable Waterways

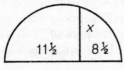

Fig. 14.14

$11\frac{1}{2} \times 8\frac{1}{2} = x^2$ which gives $x = 9\cdot88$ feet.

Triangulation You may arrange the two adjacent sides of T and t to fit in two ways, and the combination for two triangles is six shapes.

Fig. 14.15

For each of the above shapes you may arrange a third triangle in two ways adjacent to each of the four sides of the figure.

This gives eight shapes based on each of the two above figures, but the matter of duplication has to be considered.

Packaging For minimum area the box should be near cubical. The total volume of the cartons is 7680 cubic inches and the cube root of this lies between 19 and 20. But the cartons have to fit into the box, so experiment with dimensions near 20 inches. The cartons are 10 inches tall, so one side may be 20 inches long by laying two packets together. Hence we arrange them in layers of four, measuring 20 inches by 16 inches. The difficulties of fitting the cartons in the box become clear in the second part. The total volume is now 8000 cubic inches, which is the volume of a cubical box with 20 inch sides. The problem then reduces to one of arranging 8-inch by 2-inch rectangles in a square of side 20 inches, but unfortunately a four-inch square remainder always appears in our arrangement; this is the area of a carton end but not the shape. The 25 tops will have to be arranged in five rows of five.

Double Inscription 1 (1) Area of innermost square is always half of intermediate square. (2) Suppose outer square was 3 × 3 inches, its area would be 9 in². But area of each triangle (see Fig. 14.16) is 1, so area of intermediate square is 5 in².

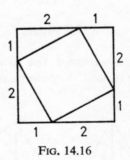

FIG. 14.16

Repeat with an outer square of side $(1 + n)$ inches. For last part, inner square is no longer half, but same ratio as other two squares.

Double Inscription 2 Imagine the inner square rotated.

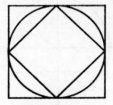

FIG. 14.17

Double Inscription 3 Draw a diagonal. Diameter of inner circle is same as length of side of square.

Cubism 1 By rotating the inner cube you may find the length of its diagonal (2 inches). Now imagine the cube of side a inches. Then by Pythagoras $a^2 + 2a^2 = 4$

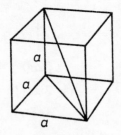

FIG. 14.18

Cubism 2 Diagonal of cube here is 2 inches. Hence from *Cubism 1*, cube is of side 1·15 inches.

Drawing Exercise 1 Draw the line 12 inches long and draw a line perpendicular at one end and 24 inches long. Now use the method outlined in the text.

The Hole in the Sphere (Solution by the use of calculus). Imagine the sphere formed by the generating curve $x^2 + y^2 = r^2$. Let the radius of the required hole be R, and the distance from the origin to where the hole cuts the sphere have an x-coordinate of a, as shown.

M

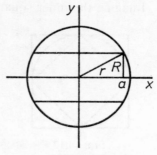

FIG. 14.19

We obtain the volume of the solid remaining by finding the volume of the sphere enclosed between the limits of $+a$ and $-a$, and subtracting the volume of a cylinder of radius R and length $2a$. Hence

$$\text{volume of solid remaining} = \pi \int_{-a}^{a} (r^2 - x^2) \, dx - 2\pi R^2 a$$

$$= \pi \left[r^2 x - \tfrac{1}{3} x^3 \right]_{-a}^{a} - 2\pi R^2 a$$

$$= 2\pi a (r^2 - \tfrac{1}{3} a^2 - r^2 + a^2)$$

$$= 2\pi a \, (\tfrac{2}{3} a^2)$$

$$= 4\pi a^3 / 3$$

Since this is half the total volume,

$$\frac{2\pi r^3}{3} = \frac{4\pi a^3}{3}$$

or

$$2a^3 = r^3$$

Since $R^2 = r^2 - a^2$, R may now be found in terms of r by simple substitution and gives an approximate solution of $R = 0 \cdot 61 r$.

Another Hole   As the diameter of the sphere is not given it would seem that this has no bearing on the result. Why not take the sphere to be of height six inches, then?

A Final Wangle   Join QR and consider the triangle PQR.

CHAPTER 9

Night Train   Number the statements 1 to 7. No. 2 indicates Edwards is not the driver; 3 & 4 Brown is not the driver; 6 & 7 Thomson is the fireman. This gives the following array:

	Brown	Thomson	Edwards	Robinson
ED	0		0	
F		1		
G				
SCA				

Complete the columns with 0s and 1s, then re-examine the statements.

Partners   Use two tables, one for 'dancing partners' and one for 'married couples'. From 'Alan dancing with Ann', and the fact that John is mentioned by name, Mary's husband and Ann's husband can not be these two. So Ann and Mary must be married to one or other of Eric and Bill. But Bill's wife is dancing with John. This now gives tables as shown:

	Dancing				Marriage			
	Joan	Ann	Mary	Pat	Joan	Ann	Mary	Pat
Alan	0	1	0	0		0	0	
John		0				0	0	
Eric		0				1		
Bill		0				0		

Professional Problem   Accountant does Smithers accounts; Johnson has a surgery (doctor or dentist); Johnson and accountant settle each other's bills and accountant gets his medical treatment free. These give the following:

	Bennett	Johnson	Smithers	Woods
Accountant		0	0	
Solicitor		0		
Doctor		0		
Dentist				

Copycats   Respective statements give

$$B'D+BD', \quad B'C+BC', \quad A'B+AB', \quad D'+D$$

Connect and simplify.

Committee Quorum   Statements give

$$J+S+M+R, \quad J'+M,$$
$$(MR)'+S', \quad (MS)'+J, \quad R'+J$$

Connect and simplify

Club Run   Statements give

$$K'+L', \quad K+L, \quad M'+K,$$
$$KL+M+N, \quad N'+M$$

Illuminations

    1. $W'+BA$ 2. $A'+R$ 3. $B+W$
    4. $BG+B'G'$ 5. $RG'+GR'$

Coach Trip

$$A'+C, \quad (AE)'+B, \quad (B+C)'+D,$$
$$D'+(BC'+B'C), \quad DE$$

Fibs and Trues   Consider Paul's reply:

(a) If he was a True, he would say he was a True.
(b) If he was a Fib, he would say . . . ? (remember he always lies)

Now you should be able to say what Pat and John are.

Abstract Art   Tabulate as follows

A	B	C	D
X		Y'	
	Y'		Z
		X/Y	W'
		Z	W

CHAPTER 10

Arithmetic Progression

(a) $n = 25$, $a = 2$, $d = 2$
(b) $n = 33$, $a = 102$, $d = 3$.

Space Problem

Lines	Spaces
3	1
4	3
5	6
6	10	and so on.

Pyramid Numbers   Base of the nth layer consists of a triangle with $\frac{1}{2}n(n+1)$ in it.

Street Plan   Approach this by considering the number of ways of reaching each intersection

A	1	1	1	1	1
1	2	3	4	5	6
1	3	6	10	15	
1	4	10	20		
1	5	15			
1	6				

The method of finding the number of ways by adding the two numbers at the adjacent corners of the square will be evident. Notice also the similarity with Pascal's triangle. The last part of the question may be difficult for people with no knowledge of the Binomial Theorem, but the number of ways of reaching the intersection of the pth and qth streets emerges as the coefficient of $a^p b^q$ in the expansion of $(a+b)^{p+q}$.

Across the Chessboard

(a) Each square may be approached from the one above or the one adjacent, so we may progressively insert the number of ways of reaching any square by adding the numbers in each of the foregoing squares, e.g.

	1	1	1	1	1
1	2	3	4	5	6
1	3	6	10	15	
1	4	10	20		
1	5	15			

and so on.

The pattern will be immediately recognized.

(b) The problem when diagonal moves are also allowed may be attempted in a similar way, but in this case remember to add the numbers above, to the left, and the one between these two, since any square may be approached in these three directions.

$$
\begin{array}{ccccc}
1 & 1 & 1 & 1 & 1 \\
1 & 3 & 5 & 7 & \\
1 & 5 & 13 & & \\
1 & 7 & & &
\end{array}
$$

and so on.

Split the Circle   Each semi-circle is a quarter of the area of the preceding semi-circle. Hence one part of the circle is of area

$$A + \frac{1}{4} A - \frac{1}{16} A + \frac{1}{64} A - \ldots$$

and the other is of area

$$A - \frac{1}{4} A + \frac{1}{16} A - \frac{1}{64} A + \ldots$$

where A is half the area of the circle. These two series may be summed to infinity to give respectively

$$A + \frac{\frac{1}{4}A}{1 + \frac{1}{4}}, \quad \frac{A}{1 + \frac{1}{4}}$$

Varied Descent   This problem is the same as *Street Plan*, but in three dimensions. Apply the same method (see Fig. 14.20).

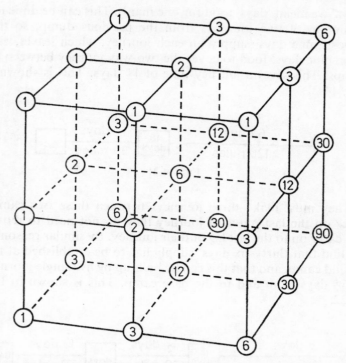

FIG. 14.20. The numbers indicate the number of different ways of reaching that point.

The Expedition The first man's food suffices for all five men for the first day and his own return on the second day.

The second man's food suffices for the remaining four men for the second day and his own return on the third and fourth days.

The third man's food suffices for the remaining three men for the third day and his own return over a three-day period.

And so on.

Third Expedition Starting from the end of the problem, it is clear that the men do the last 120 miles in a direct run since they can just carry the amount of food required for this distance. Hence the last food dump must be 120 miles from the end and they must establish eighteen days' food here (throughout by 'days' we mean 'days' food for one man'). This can be done in a minimum of two journeys from the previous dump, so they deposit nine days' supply on each journey, which leaves, each time, nine days' food to cover the two-way journey between the camps. This gives a journey time of $1\frac{1}{2}$ days. This is shown in Fig. 14.21.

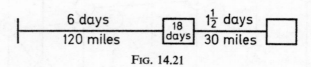

FIG. 14.21

They must make three journeys between these two camps, since on the last outward journey they do not need to return, but carry on to the last leg of their journey. By similar reasoning we find that thirty-six days' supply has to be established at this second camp, and that this can be built up by five single journeys of $1\frac{1}{2}$ days from and to the base camp. This is shown in Fig. 14.22.

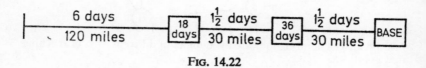

FIG. 14.22

Strictly Numerical 1 Suppose the numbers are a/r^2, a/r, a, ar and ar^2. Their product is a^5.

The general result may be obtained similarly.

Strictly Numerical 2 Let a be any term and r the common ratio. Then the sum to infinity of the terms after a is $ar/(1-r)$. If

$$a > ar/(1-r)$$

then

$$a - ar > ar$$
$$a > 2ar$$
$$r < \tfrac{1}{2}$$

Strictly Numerical 3 Let a be the first term in each case. If d

represents the common ratio (or difference), the *AP* is

$$a, a+d, a+2d$$

and the *GP* is

$$a, ad, ad^2$$

But

$$ad^2 = d(a+2d)$$
$$ad^2 = ad+2d^2$$
$$a(d^2-d) = 2d^2$$

Hence

$$a = \frac{2d^2}{a^2-d} \quad \text{or} \quad a = \frac{2d}{d-1}$$

Now find possible values for *a* and *d*. Why are there only two possible solutions to the problem?

CHAPTER 11

European Tour

(a) Consider, for example, Portugal and Spain (P,S), Denmark and West Germany (D, WG), and Norway and Sweden (N,S)

(b) Consider number of boundaries that each country has with adjacent countries, e.g.

Portugal (P)	1
Spain (S)	2
France (F)	6
Italy (I)	4
Greece (G)	3
Denmark (D)	1
Poland (P)	3

Continuous Line Count the number of lines meeting at each intersection.

All Around the House 1 Represent the rooms by points and show how they may be joined in a continuous line.

CHAPTER 13

Long Division The main clues are that the middle digit of the divisor gives 0 and 4 in the second and last partial products and similarly for the first digit of the divisor which gives 0 and 5. Consider also that the first two partial products are 4-digit numbers and the last only 3-digit numbers.

Multiplication 1 The fact the $T \times T$ gives T means that T is 1 or 5 or 6. Notice also that $H \times T$ gives P and so does $O \times T$, while $S + P$ gives S. The fact that each product has four digits means that H cannot be less than 3.

Multiplication 2 Points to note: $F \times H = H$, $M \times H = B$, $M + C = B$ and $M + J = M$. The middle line only gives five digits.

A Lot of Subtraction $A - J = A$ gives the vital clue.

Letter Division E multiplied by C, B or A gives E. Only two-digit partial products are involved. (A simple problem, but the solution is not unique.)

Answers

CHAPTER 1

Work This Out   Further solutions are

```
6 2 1 8      5 6 1 2
5 9 4 3      4 9 7 8
───────      ───────
  2 7 5        6 3 4
```

Other solutions are possible.

Relatives   Second (or quarter) cousins.

Suspects   Thomson

Pupil's Choice   Physics/Biology; Chemistry/Physics; Chemistry/Biology; Chemistry/Eng. Science; Physics/Gen. Science; Biology/Physics-with-Chemistry; Biology/Eng. Science.

Board Meeting   Bell/Hartley; Archer/Foulds; Earnshaw/Chivers; Dakin/Gerrard.

Key Situation   Twisted twice.
None (Yale-type locks have to be returned to the normal position before the key can be removed).

Carbon Copy   Top left-hand corner of each side—right way up. Bottom right-hand corner of each side—inverted.

CHAPTER 2
Advanced Warning   Four.

The Present   322; 87.

Bases and Powers   121; 1331.

Elevenses   See 'Hints for Solutions'.

Duo-Decimals   1118, 561, 1294.

CHAPTER 3

Digital Sum 1

```
        13 — 2 — 6
       5   \ | /   4
      3 ——— 7 ——— 11
      10   / | \   9
        8 —12— 1
```

Digital Sum 2

```
            12
    11 — 3 — 4 — 8
        6       1
    5 — 2 — 10 — 9
            7
```

Digital Sum 3

```
          2
       9  |  7
    4 ——— 5 ——— 6
       3  |  1
          8
```

Odd Sums 1^3, 2^3, 3^3, 4^3, etc. For explanation see 'Hints'.

Triangular Tesselation 1   Two: rhombuses and regular hexagons

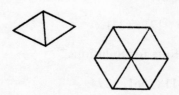

ANSWERS   189

Quadrilaterals   Yes, as shown below:

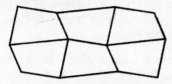

Tiling Shapes   (a) No (b) Yes (c) No (d) No.   The sum of the angles meeting at a point must equal 360°.

Multi-Shaped Tesselations   All are possible—see 'Hints for Solutions'.

CHAPTER 4

Century Ahead $0+1+2+3+4+5+6+7+8\times9$

John and Mary   John is at present 20 years old.

Joan and Bill   Bill is 28 and Joan is 21.

Trenching   Seven times as deep.

Fly Catcher   120 miles.

Birthday Party   13, 20, 26.

Procession   6 m.p.h.

Worth a Note   14.

The Narrow Bridge   10 m.p.h.

Right-Angled Triangles 2   16, 63, 65
                            33, 56, 65.

Baling Out   He started with 80 and finished with 15.

Sum and Product   11, 12, 15, 24, 36

Small Change   9.

Food Mixer   5 ounces of *A* with 12 ounces of *B*.

Oil Production   (a) 4000 light; 2000 heavy   (b) 4250 heavy; 1750 light.

CHAPTER 5

First   The rope moves so that the monkey and the weight are always level with each other.

Second It falls. By the Principle of Archimedes, when the iron is floating it displaces its own weight of water; when it sinks it displaces its own volume. Iron is denser than water, so the first displacement is greater.

Third When motion takes place in a string supporting a weight, the tension becomes less if the motion is in the direction of the weight. Hence when the 50g weight falls, the tension in the string around its pulley decreases, so that the tension in the string around the top pulley decreases on that side. Consequently the weights which are still fastened together move downwards.

Fourth To a certain extent it is sound. The total weight on the bridge will still be the same, but the weight will be distributed over a large area instead of being concentrated on the points where the wheels are in contact with the bridge. But much depends on the individual bridge. The stresses a bridge can take depend on (a) stress on particular points, and (b) overall loading. The relative importance of (a) and (b) vary in different bridges. This transporter only assists in (a).

Fifth In the centre of the cut cube is a small cube whose six faces have all to be cut by the saw. It is impossible to cut these in less than six saw cuts.

Sixth The logic is correct; check the statement!

CHAPTER 6

Motorist's Choice 120.

Light Reading 14, 22.

The College Students 20%.

Market Survey You either demand a fresh survey or dismiss them on the grounds that their figures do not balance.

Language Laboratory 16.

Through Trains False.

Top People True.

Office Routine Only brunettes can do shorthand in this office. (Other conclusions are possible.)

Animal Crackers It may or may not be true; insufficient information is given.

Committee Membership Membership of the Refreshment Committee or of the Sports Committee excludes one from membership of the General Committee.

CHAPTER 7

Navigable Waterways Yes.

Triangulation 6, 12, 20, etc. $n(n+1)$ where n is the number of triangles.

Packaging
   (i) 20 inches by 16 inches by 24 inches.
   (ii) 40 inches by 20 inches by 10 inches.

Double Inscription 1
One inch.
Ratio of areas of squares is 5/18
Ratio of areas $= (1+n^2)/2(1+n)^2$
Ratio of areas $= (1+n^2)^2/(1+n)^4$

Double Inscription 2 Area of 2 square inches, i.e. of side $\sqrt{2}$ inches.

Double Inscription 3 $\sqrt{2}$ inches.

Cubism 1 1·15 inches.

Cubism 2 1·15 inches.

Drawing Exercise 1

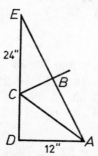

CB is perpendicular bisector of AE.
ACD is required triangle.

Drawing Exercise 2 Draw the base *AB* and a line *PQ* parallel to it at the given altitude away. Bisect *AB* at *C*. Suppose the length of the perimeter less the base is *XY*, with mid-point *Z*, then produce *AB* to *D*, so that *CD* = *XZ*. Draw *CE* perpendicular to *AB* and make *BE* = *CD*. With centre *C* make an arc, radius *CE* to cut *PQ* in *M*. Draw another arc, centre *C*, radius *CD*, and let it cut *CM* produced in *N*. Now draw *NR* perpendicular to *PQ* and join *AR*, *BR* to form the required triangle *ARB*.

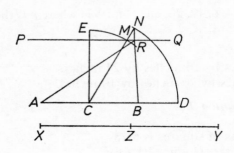

Another Hole   36π cubic inches.

A Final (W)angle   60°.

CHAPTER 8

Cross-number Puzzle 1
Across   1 **17**   3 **62**   5 **5832**   9 **11**   11 **42**   12 **24**   13 **27**
15 **13**   16 **48**   17 **64**   18 **18**   19 **22**   21 **39**   23 **9801**
25 **16**   26 **65**
Down   2 **75**   3 **62**   4 **31**   6 **84**   7 **32**   8 **64**   10 **1248**
12 **2343**   14 **78**   15 **16**   18 **13**   19 **28**   20 **20**   22 **97**
23 **96**   24 **16**

Cross-number Puzzle 2
Across   1 **147**   3 **369**   5 **880**   7 **49**   9 **19**   10 **81**   11 **22**
12 **24**   14 **56**   16 **607**   18 **512**   19 **250**
Down   1 **164**   2 **78**   3 **30**   4 **999**   6 **88**   8 **984**   9 **125**
12 **275**   13 **90**   15 **640**   16 **62**   17 **72**

Cross-number Puzzle 3
Across   1 **531**   3 **108**   6 **52**   7 **1004**   9 **28**   10 **147**
12 **502**   14 **666**   15 **54**   16 **32**   17 **456**   19 **511**   22 **143**
24 **22**   25 **1728**   28 **41**   29 **270**   30 **578**

Down 1 **5280** 2 **111** 3 **117** 4 **00** 5 **80** 6 **525** 8 **456**
11 **49** 13 **256** 14 **625** 17 **401** 18 **34** 20 **1248** 21 **121**
22 **180** 23 **385** 26 **72** 27 **27**

Cross-number Puzzle 4
Across 1 **1066** 3 **112** 5 **30** 6 **1357** 7 **246** 8 **624**
11 **440** 12 **169** 13 **1760** 15 **22** 17 **007** 18 **3600**
Down 1 **17** 2 **63360** 3 **10** 4 **24** 5 **37** 6 **144** 7 **24**
8 **61616** 9 **260** 10 **49** 13 **12** 14 **20** 15 **27** 16 **60**

CHAPTER 9

Wotzits
  Grommit, sneezle, yatter, tamble
  Grommit, sneezle and nosher
  Sneezle, nosher and tamble.

Night Train Driver: Robinson; Fireman: Thomson; Guard: Brown; Sleeping car attendant: Edwards.

Partners Alan and Pat; John and Joan; Eric and Ann; Bill and Mary.

Professional Problem Accountant: Woods; solicitor: Bennett; doctor: Smithers; dentist: Johnson.

Copycats Bill.

Committee Quorum Macmillan; Smith; Macmillan and Jones; Macmillan, Jones and Smith; Macmillan, Jones and Riley.

Club Run Ken and Mary, or Ken, Mary and Nancy.

Illuminations Blue, green and white.

Coach Trip Dickle and Egham with either Brabham or Coppleton or both.

Fibs and Trues Pat: True; John: Fib.

Abstract Art $A = X, B = Y, C = Z, D = W$

Hats Off The answer to all the questions is, yes, it is possible.

CHAPTER 10

Arithmetic Progression 650, 4950.

Circle Intersections 2, 6, 12, . . . , $n(n-1)$.

N

MATHEMATICAL PUZZLES AND PERPLEXITIES

The Space Problem $\frac{1}{2}(n-2)(n-1)$.

Pyramid Numbers   Successive addition of 1, 3, 5, 7, 9 . . . gives square numbers. nth term of pyramid series is $\frac{1}{6}n(n+1)(n+2)$.

Figurate Numbers   The linear, plane, pyramid numbers are obtained as successive oblique columns of the Pascal triangle.

Street Plan   462; $(p+q)!/q!p!$ where 4! means 4.3.2.1.

Across the Chessboard   3432; 48639.

Skew Pascal Triangle   The Fibonacci series is obtained.

Cube It   The sum of the cubes is the square of the sum of the integers, e.g. $1^3+2^3+3^3+4^3 = (1+2+3+4)^2$. The explanation comes from the proof of the general case

$$4n^3 = \{(n+1)n\}^2 - \{(n-1)n\}^2$$
$$4(n-1)^3 = \{(n-1)n\}^2 - \{(n-2)(n-1)\}^2$$

and so on to

$$4.2^3 = (2.3)^2 - (1.2)^2$$
$$4.1^3 = (1.2)^2 - (0.1)^2$$

Adding each side we obtain

$$4(1^3+2^3+3^3+ \ldots +n^3) = n^2(n+1)^2$$

Split the Circle   3:2.

Varied Descent   90.

An Economical Ruler   Various possible solutions: 1, 4, 5, 10; 2, 5, 8, 11; 1, 2, 6, 9.

The Expedition   (a) 10 days   (b) $2nd/(n+1)$ days.

Third Expedition   18 days.

Strictly Numerical 1   See 'Hints for solution'.

Strictly Numerical 2   See 'Hints'.

Strictly Numerical 3   4, 6, 8; 4, 8, 16   or   3, 6, 9; 3, 9, 27.

Geometric Growth   Nine days.

CHAPTER 11
Venn Diagrams

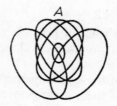

The inner small ellipse is regarded as *outside* ellipse *A*.

European Tour   (a) No.   (b) No.

Continuous Line   (a) Unicursal, not closed; (b) not unicursal; (c) unicursal, not closed; (d) unicursal, closed; (e) unicursal, closed; (f) unicursal, not closed.

Letter Problem   B C D I J L M N O P R S U V W Z.

All Around the House 2   There are an even number of doors in each room.

The Federation 1   An example is

 or

Ring-Ring

String-Ring   One way is

Moebius Strip   The first is a one-sided, one-edged solid.

Deformations   (a), (b), (d) and (h) are transformable, also (c) and (g).

Round the Cone   An ellipse.

N*

Spider Circuit A regular hexagon. (This may be demonstrated by putting a rubber band around a cube.)

CHAPTER 12

First The total of possible arrangements is 24. (The first letter may be put in four envelopes, the next in three and so on—total $4 \times 3 \times 2 \times 1$).

Number of these arrangements with all letters correct = 1
Number of these arrangements with two letters correct = 6
Number of these arrangements with one letter correct = 8
Hence, number of ways with all letters being wrong = $24 - 15 = 9$.

Second 802. When arranged in order on a bookshelf the pages are in reverse order to the volumes. So the bookworm only eats through the two *complete* volumes II and III. (See diagram.)

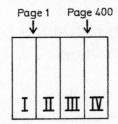

Page 1   Page 400

I   II   III   IV

Third As long as he sells nine by taking five of the cheap ones and four of the dear ones all is well, but after thirty-six transactions he has exhausted his supply of cheap oranges and is then selling nine of the dearer oranges at a price below their value. This is really an example on averages, and the important principle is that one cannot average averages. Harry should have taken the total price expected (405p) and divided this by the total oranges in order to obtain a selling price, e.g. 9p for eight.

Fourth The pattern is always 1 4 9 6 5 6 9 4 1 and the repetition every ten is explained simply by the fact that the units digit of the squares is always formed from the product with itself of the units digit of the original number. Part (b) is explained as follows: Take $4 \times 4 = 16$ as the example.

$$6 = 10 - 4$$

So

$$6 \times 6 = (10 - 4)(10 - 4)$$
$$= 100 - 80 + 16$$

Hence the units digit is formed by squaring four, and similarly for the others; for example

$$3 \times 3 = 9$$
$$7 \times 7 = (10-3)(10-3)$$
$$= 100-60+9.$$

Fifth The explanation depends on the time between trains in the two directions. If there is an interval of A minutes between a train to Dench and one to Parrock, and an interval of B minutes between the one to Parrock and the next one to Dench, then if $A = B$, Len would have an equal chance of seeing either train first. But if B is greater than A, the chance of a Dench train coming first is greater than that of the Parrock train. The trouble lies in Len going at any random time. The clock diagram illustrates

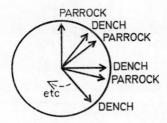

Sixth In mathematics the multiplication of two negative numbers or two positive numbers gives a positive number. In logic, two negatives together, or two affirmatives together, give an affirmative. The question to ask is one requiring two successive replies which are then bound to give the truth. An example is 'What would you reply if I asked you if Pat lived here?'

CHAPTER 13
Long Division

```
215) 123195 ( 573
     1075
     ————
      1569
      1505
     ————
       645
       645
```

Multiplication 1

1	2	3	4	5	6	7	8	9	0
M	S	L	A	T	H	–	O	Z	P

Multiplication 2

1	2	3	4	5	6	7	8	9	0
M	K	F	A	H	G	C	J	B	–

A Lot of Subtraction

```
9 8 7 6 5
4 3 2 1 0
─────────
5 5 5 5 5
```

The units column is fixed, but the other columns are not, so it is possible to arrange these in any order. There are twenty-four possible arrangements.

Letter Division

$$6420 \div 20 = 321 \quad \text{or} \quad 6930 \div 30 = 231$$

On All Fours

$$31666 \div 142 = 223$$

A Fourth-Order Magic Square   Various possibilities, e.g.

1	15	14	4
12	6	7	9
8	10	11	5
13	3	2	16

1	12	13	8
15	6	3	10
4	9	16	5
14	7	2	11

1	14	15	4
8	11	10	5
12	7	6	9
13	2	3	16

Diagonal Latin Squares   Example

1	2	3	4
4	3	2	1
2	1	4	3
3	4	1	2

Not possible with 3×3.

Modulo 5 Composition

```
4 | 4 0 1 2 3
3 | 3 4 0 1 2
2 | 2 3 4 0 1
1 | 1 2 3 4 0
0 | 0 1 2 4 3
--+----------
+ | 0 1 2 3 4
```

```
4 | 0 4 3 2 1
3 | 0 3 1 4 2
2 | 0 2 4 1 3
1 | 0 1 2 3 4
0 | 0 0 0 0 0
--+----------
× | 0 1 2 3 4
```

Modulo 5 Addition

```
4 | 1 3 0 2 4
3 | 0 2 4 1 3
2 | 4 1 3 0 2
1 | 3 0 2 4 1
0 | 2 4 1 3 0
--+----------
+ | 2 4 1 3 0
```

A diagonal Latin Square

Magic Cube

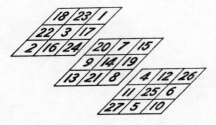

Elements of Modulo 8

7	7	5	3	1
5	5	7	1	3
3	3	1	7	5
1	1	3	5	7
	1	3	5	7

A Latin Square (not diagonal).

Switching Table

D	D	C	B	A
C	C	D	A	B
B	B	A	D	C
A	A	B	C	D
	A	B	C	D

A Latin Square.

This is identical with the previous example in its arrangement, although the initial problem was entirely different. The reason for this borders on the idea of groups and operations within them. The interested reader will find further detail in an elementary book on groups.

Bibliography

Puzzle Books

There is a wide variety of these, some general in nature, others covering particular types of puzzle. Worthy of particular mention are:

W. W. ROUSE BALL, *Mathematical Recreations and Essays*, Macmillan.

M. KRAITCHIK, *Mathematical Recreations*, Allen & Unwin.

J. NORTHROP, *Puzzles and Paradoxes in Mathematics*, Penguin.

C. R. WYLIE, 101 *Puzzles in Thought and Logic*, Dover.

M. GARDNER, *Mathematical Puzzles and Diversions from Scientific American*, Bell; *More Mathematical Puzzles and Diversions from Scientific American*, Bell; *New Mathematical Diversions*, Allen & Unwin; *Further Mathematical Diversions*, Allen & Unwin.

References on Mathematical Topics

There are many books which provide the necessary mathematics mentioned in the present book. The following include some interesting general books on mathematics as well as some fairly elementary texts; they are not books of puzzles.

The algebra topics (series, theory of equations, etc.) will be found in any book of about sixth form standard, e.g. TUCKEY AND ARMSTRONG, *Algebra*, Longmans.

For Pythagorean Triples see Kraitchik (above), and for more work on magic squares see the same book and Rouse Ball (also above).

Tesselations and space filling are to be found in:

H. STEINHAUS, *Mathematical Snapshots*, Oxford University Press

H. WEIL, *Symmetry*, Princeton University Press.

For elementary work on sets and Boolean algebra see:

C. A. R. BAILEY, *Sets and Logic*, Books I and II, Edward Arnold.

A. P. BOWRAN, *A Boolean Algebra*, Macmillan.

Some of the uses of the Fibonacci series are covered in:

F. LAND, *The Language of Mathematics*, John Murray.

Some of the newer topics are introduced in the first of the following books—including groups (see answers to Chapter 13). The second book is one of general interest to 'dabblers' in mathematics:

W. W. SAWYER, *Prelude to Mathematics*; *Mathematician's Delight*, Penguin.

Finally a book with a self-explanatory title:

R. C. READ, *Tangrams*, Dover.

GEORGE ALLEN & UNWIN LTD

Head office:
40 Museum Street, London, W.C.1
Telephone: 01–405 8577

Sales, Distribution and Accounts Departments
Park Lane, Hemel Hempstead, Herts.
Telephone: 0442 3244

Athens: 7 Stadiou Street, Athens 125
Barbados: Rockley New Road, St. Lawrence 4
Bombay: 103/5 Fort Street, Bombay 1
Calcutta: 285J Bepin Behari Street, Ganguli, Calcutta 12
Dacca: Alico Building, 18 Motijheel, Dacca 2
Hornsby, N.S.W.: Cnr. Bridge Road and Jersey Street, 2077
Ibadan: P.O. Box 62
Johannesburg: P.O. Box 23134, Joubert Park
Karachi: Karachi Chambers, McLeod Road, Karachi 2
Lahore: 22 Falettis' Hotel, Egerton Road
Madras: 2/18 Mount Road, Madras 2
Manila: P.O. Box 157, Quezon City, D-502
Mexico: Serapio Rendon 125, Mexico 4, D.F.
Nairobi: P.O. Box 30583
New Delhi: 4/21–22B Asaf Ali Road, New Delhi 1
Ontario, 2330 Midland Avenue, Agincourt
Singapore: 248C–6 Orchard Road, Singapore 9
Tokyo: C.P.O. Box 1728, Tokyo 100–91
Wellington: P.O. Box 1467, Wellington